Praise for *Toxic Legacy*

"*Toxic Legacy* is a bold and heroic work that reveals how today's most well-respected science confirms the existential threat posed by the herbicide glyphosate. Dr. Seneff courageously defends her position in the face of seemingly insurmountable opposition from industry at every turn. And for this we owe her incalculable gratitude. *Toxic Legacy* will stand shoulder to shoulder with Rachel Carson's *Silent Spring*, not just defining the pervasive threat to us and future generations, but more importantly, articulating what we can do right now to change our destiny. Unquestionably, one of the most important books of our time."

—David Perlmutter, MD, #1 *New York Times* bestselling author of *Grain Brain* and *Brain Wash*

"A highly readable, fascinating look at how man-made chemicals invade the air we breathe, the water we drink, and the food we eat. *Toxic Legacy* is both a scientific exposé and an inspiring call to action. In revealing the diabolical mechanisms by which glyphosate damages human health, Dr. Seneff's work will change the way we all think about food."

—Mark Hyman, MD, founder and director, The UltraWellness Center; Head of Strategy and Innovation, Cleveland Clinic Center for Functional Medicine; and a thirteen-time *New York Times* bestselling author

"Anyone who wants to understand a root cause of the massive epidemics of chronic diseases we are experiencing—from autism to non-alcoholic fatty liver and 30+ other devastating diseases rampant in today's society—can find the answers in *Toxic Legacy*. Dr. Seneff is to be complimented in her comprehensive, in-depth connection of the dots from glyphosate through the myriad biochemical and physiological processes altered, to the tragic consequences from the indiscriminate application of the Roundup and other glyphosate-based herbicides. The complex chemical perturbations are presented in an easily-understood manner, and the systems are well documented. Her 'relaxed' writing style makes for easy reading and ready comprehension of this important information."

—Don M. Huber, professor emeritus, Purdue University and retired colonel, US Army, Medical Intel

"Dr. Seneff is a senior scientist at MIT. She not only understands molecular biology at a deep level but also has the unique ability to translate extremely complex technical concepts into easy-to-understand language. This is a must-read book to help you comprehend one of the most significant toxic threats unleashed on the world. *Toxic Legacy* is the modern-day equivalent of Rachel Carson's *Silent Spring* from 1962—one of the greatest science books of all time."

—Dr. Joseph Mercola, founder, mercola.com, the most visited natural health site for eighteen years

"Glyphosate (the active ingredient in Roundup) and the other toxic chemicals and GMOs of industrial agriculture are the primary drivers of the chronic disease epidemic that has degenerated public health and the environment. Until we drive these poisons off the market and make the transition to organic and regenerative farming practices, our health, the health of our children, and the health of the planet are at risk. Glyphosate is the DDT of the twenty-first century, and *Toxic Legacy* is essential reading for everyone who cares about food and health."

—Ronnie Cummins, Organic Consumers Association; author of *Grassroots Rising*

"At last, a scientist with impeccable credentials has painstakingly assembled, categorized, and presented the growing body of evidence that highlights the negative impacts of the most widely used pesticide in the world.

"Stephanie's forensic analysis demonstrates beyond reasonable doubt that the use of Roundup and its active ingredient glyphosate is not only poisoning the planet but also its human population.

"Everyone should read this book. It's about our future health and the health of the planet. What issue could be more important than that? As an organization that has been highlighting these potential harms for more than a decade, The Sustainable Food Trust hopes that this milestone publication gets the publicity it needs and deserves."

—Patrick Holden, founder and chief executive, Sustainable Food Trust

"Seneff takes us on a shocking biochemical journey through the deleterious effects of glyphosate on the environment and humans. The author clearly explains the ever-growing body of scientific evidence of the insidious consequences of its continued, massive application across the world. . . . Seneff is precise about the biochemistry involved, but she is a genial, attentive guide. . . . The two most salient—and devastating—points that Seneff highlights: First, glyphosate, which shows up in our soil, water, and even air, is disturbingly pervasive . . . ubiquitous . . . 'nearly impossible for even the most diligent person to avoid.' Second, the agricultural industry, taking a page from the tobacco industry's playbook, does everything it can to hide the dangers: Monsanto and other companies censor research and proliferate junk science, raking in profits by turning a blind eye to the chronic illnesses resulting from glyphosate use. Comparisons will be made to Rachel Carson's *Silent Spring*—and they should be. We can only hope Seneff's work goes on to rival Carson's in reach and impact. A game-changer that we would be foolish to ignore."

—KIRKUS REVIEWS (starred)

"Stephanie Seneff's *Toxic Legacy* continues on the path laid by Rachel Carson in *Silent Spring*. Seneff provides the scientific evidence of how, by disrupting our bodies' metabolic pathways, glyphosate is at the root of multiple metabolic, neurological, autoimmune diseases that have taken epidemic form. More significantly, Seneff's book introduces us to the complex, sophisticated, metabolic processes of life, from the soil to our gut microbiome. It is a book for governments who want to ban glyphosate. It is a book for every citizen who seeks to regenerate the health of the planet and people."

—DR. VANDANA SHIVA, director, Navdanya;
coauthor of *Oneness vs. the 1%*

"There is nothing more important than the health of our current and future generations. There is nothing more costly than jeopardizing the physical and mental health of generations of our populations. Without our health, we cannot effectively combat climate change, homelessness, poverty, racial injustice, addiction, or other daunting issues. This book unlocks why we should and how we can resolve a myriad of skyrocketing health issues plaguing society. If heeded, the result of banning glyphosate would put us in the direction of recovering our health and the future of human life on this planet."

—ZEN HONEYCUTT, founding executive director, Moms Across America

"Dr. Stephanie Seneff, established and out-of-the-box critical thinker / researcher, lays down the foundation of our understanding of the toxicity from glyphosate, in a clear and concise format, from the practical to the molecular mechanisms on how glyphosate does harm. Although Dr. Seneff leans heavily on the biochemical basis of glyphosate's modes of action, her information is understandable and carefully cited for the more investigative readers. She carefully describes the blueprint for the myriad of chronic diseases now prevalent globally and how they are created around the ubiquitous glyphosate. Her book is not only an analytic reference treasure, but a call to action to denounce and reverse our self-imposed toxic legacy."

—MICHELLE PERRO, MD; coauthor of *What's Making our Children Sick?*;
executive director and cofounder, GMO Science

"For the last decade, while 'science' and 'health' journalists were gleefully content to repeat the limited corporate talking points about glyphosate's safety, Dr. Seneff was on the hunt for deeper truths. *Toxic Legacy* may be one of the most important literary journeys, weaving in a bounty of irrefutable evidence, essential science, and the personal journey of one of the most treasured scientific researchers of our time."

—JEFFEREY JAXEN, investigative health reporter

"Glyphosate has become a bit like sugar, used by everyone for everything, everywhere. It has spread, largely invisibly, across farms and gardens and parks and so, too, into our food, bodies, and lives. This book lays out, in expert technical detail by a scientist with huge and broad knowledge, what this ubiquitous and 'uniquely diabolical' chemical and its formulations could be doing. And how the deeply worrying, chronic consequences this may have for our health are only being recognized —from affecting our critically important gut microbes to causing hormone disruption and DNA damage. It's also an intriguing, if complex, lesson on the dense web of interconnected bodily functions, structures, enzymes, and chemicals. Glyphosate has also allowed a kind of sterilization of farming with even wider consequences for the ecosystems on which we all depend. Dr. Seneff shows clearly how we need to act fast to curb its use and so change our diets, our green spaces, and how we farm."

—VICKI HIRD, MSc, FRES, food and environmental writer
and campaigner; author of *Rebugging the Planet*

"Monsanto knew for decades that glyphosate causes cancer and a deadly retinue of other devastating illnesses. Instead of warning its customers and consumers about those risks, Monsanto manipulated the science, defrauded regulators, bribed prominent researchers, transformed the EPA pesticide division into a cesspool of corruption, and promoted propaganda worldwide, systematically lying to the public that the deadly pesticide was safe. This company injured public health, destroyed our soils, exterminated species, wiped out small farmers, and deprived the public of their fundamental civil right of informed consent. Monsanto made a special project of discrediting and destroying scientists, advocates, and reformers who exposed its corrupt cover-up. Among the most prominent of these was heroic MIT researcher Dr. Stephanie Seneff. In 2018, I was fortunate enough to be a part of the legal team that finally brought Monsanto to justice. We relied heavily on Dr. Seneff's research to achieve this victory."

—ROBERT F. KENNEDY, JR.

TOXIC LEGACY

How the Weedkiller Glyphosate Is Destroying Our Health and the Environment

STEPHANIE SENEFF, PhD

Chelsea Green Publishing
White River Junction, VT
London, UK

Project Manager: Patricia Stone
Developmental Editor: Brianne Goodspeed
Copy Editor: Diane Durrett
Proofreader: Nancy A. Crompton
Indexer: Shana Milkie
Designer: Melissa Jacobson
Page Layout: Abrah Griggs

Printed in the United States of America.
First printing June 2021.
10 9 8 7 6 5 4 3 2 21 22 23 24 25

Our Commitment to Green Publishing

Chelsea Green sees publishing as a tool for cultural change and ecological stewardship. We strive to align our book manufacturing practices with our editorial mission and to reduce the impact of our business enterprise in the environment. We print our books and catalogs on chlorine-free recycled paper, using vegetable-based inks whenever possible. This book may cost slightly more because it was printed on paper that contains recycled fiber, and we hope you'll agree that it's worth it. *Toxic Legacy* was printed on paper supplied by Sheridan that is made of recycled materials and other controlled sources.

Library of Congress Cataloging-in-Publication Data

Names: Seneff, Stephanie, author.
Title: Toxic legacy : how the weedkiller glyphosate is destroying our health and the environment / Stephanie Seneff, PhD.
Description: White River Junction, VT : Chelsea Green Publishing, [2021] | Includes bibliographical references and index.
Identifiers: LCCN 2021010815 (print) | LCCN 2021010816 (ebook) | ISBN 9781603589291 (Hardcover) | ISBN 9781603589307 (eBook)
Subjects: LCSH: Pesticides—Toxicology. | Pesticides—Environmental aspects. | Glyphosate—Health aspects.
Classification: LCC RA1270.P4 S446 2021 (print) | LCC RA1270.P4 (ebook) | DDC 363.738/498--dc23
LC record available at https://lccn.loc.gov/2021010815
LC ebook record available at https://lccn.loc.gov/2021010816

Chelsea Green Publishing
85 North Main Street, Suite 120
White River Junction, Vermont USA

Somerset House
London, UK

www.chelseagreen.com

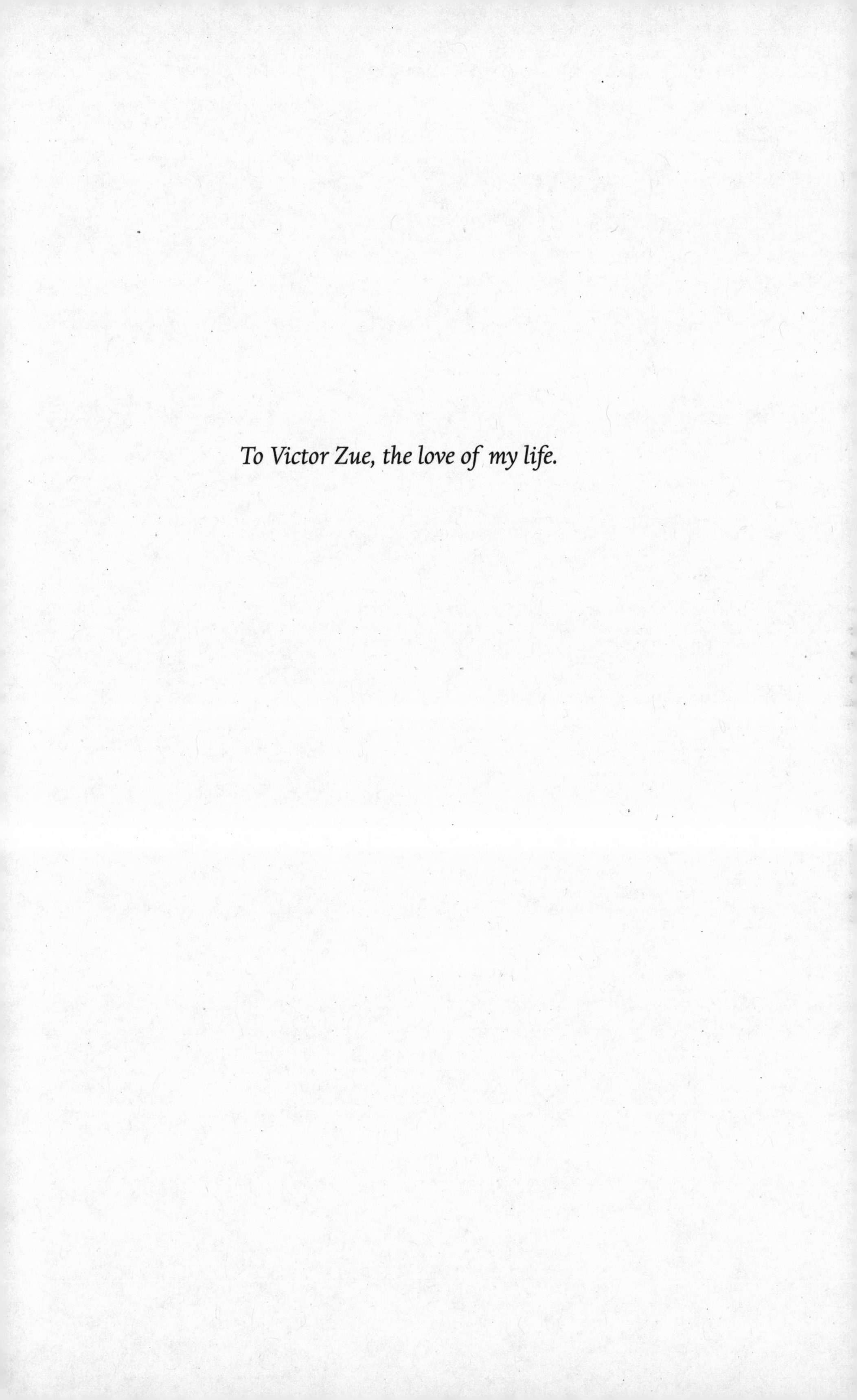

To Victor Zue, the love of my life.

Contents

Introduction

We haven't inherited this planet from our parents, we've borrowed it from our children. We have not borrowed our children's future—we have stolen it and we're still stealing it now, and it's time we get together, whatever our religion, whatever our culture, get together and start changing the way—changing our attitude—so that we can leave a better world for our children, whom we love.

—Jane Goodall

Glyphosate. Not exactly a word that rolls off the tongue. A word that was not even in my vocabulary for the first 64 years of my life. Then in September 2012, I was invited to give a talk about the dangers of statin drugs at a nutrition conference in Indianapolis. I noticed that a botanist whose agricultural research focused on the epidemiology and control of plant pathogens, Dr. Don Huber from Purdue University, was speaking on the topic of "glyphosate." Even though I didn't recognize the chemical's name, I thought it might be useful for me to find out what it was.

In the previous five years, I'd been on a dogged journey to identify environmental factors that might be causing the increase in autism among America's children. Characterized by social deficits, repetitive behaviors, and impaired cognitive abilities, autism spectrum disorder can present as relatively mild, or it can be extreme, requiring full-time lifelong care. Like many scientists, I'd noticed that the rates of autism spectrum disorders had been rising dramatically over the past few decades, in ways that could not be accounted for by diagnostic criteria. Based on a survey conducted by the United States Health Resources and Services Administration (HRSA) in 2016, the prevalence of autism in the United States is about 1 in 40 children.[1] According to the Centers for Disease Control and Prevention, the prevalence of autism among 12-year-old children is about 1 in 54, and it is four times more common among boys than among girls.[2]

By the time I attended Dr. Huber's presentation on glyphosate, I had already learned a great deal about the complicated medical conditions that often accompany autism, including a disrupted gut microbiome, inflammatory and leaky gut, nutrient malabsorption, food sensitivities, vitamin and mineral deficiencies, and impaired methylation and sulfation pathways. I had been assiduously studying the toxic effects of various metals and chemicals in the environment: mercury, fluoride, lead, aluminum, plastics, polychlorinated biphenyls, polysorbate 80, and other endocrine disruptors and carcinogens. I was also investigating the role of diet and the overuse of antibiotics. I was trying to find something in the environment that had become more pervasive in the past two decades, in step with the dramatic rise in autism rates, that might explain the diverse symptoms associated with the brain dysfunction we were seeing.

What I learned from Dr. Huber is that glyphosate is the active ingredient in the herbicide Roundup. While glyphosate isn't a household name, everyone has heard of Roundup. Drive across the United States and you'll see vast fields marked with crop labels that say "Roundup Ready." Monsanto, the Missouri-based company that was Roundup's original manufacturer, was acquired by the Germany-based company Bayer in 2018 as part of its crop science division. Monsanto has touted glyphosate as remarkably safe because its main mechanism of toxicity affects a metabolic pathway in plant cells that human cells don't possess. This is what—presumably—makes glyphosate so effective in killing plants, while—in theory, at least—leaving humans and other animals unscathed.

But as Dr. Huber pointed out to a rapt audience that day, *human cells* might not possess the shikimate pathway but almost all of *our gut microbes* do. They use the shikimate pathway, a central biological pathway in their metabolism, to synthesize tryptophan, tyrosine, and phenylalanine, three of the twenty coding amino acids that make up the proteins of our body. Precisely because human cells do not possess the shikimate pathway, we rely on our gut microbiota, along with diet, to provide these essential amino acids for us.

Perhaps even more significantly, gut microbes play an essential role in *many* aspects of human health. When glyphosate harms these microbes, they not only lose their ability to make these essential amino acids for the host, but they also become impaired in their ability to help us in all the other ways they normally support our health. Beneficial microbes are more sensitive to glyphosate, and this causes pathogens to thrive. We know, for example, that gut dysbiosis is associated with depression and other mental disorders.[3] Alterations in the distribution of microbes can cause immune dysregulation and autoimmune disease.[4] Parkinson's disease is strongly linked to a proinflammatory gut microbiome.[5] As has become clear from the remarkable research conducted on the human microbiome in the past decade or so, happy gut bacteria are essential to our health, including in ways that researchers have yet to fully understand. It's worth remembering that Roundup hit the market—and was declared safe—before much of this groundbreaking research on the human microbiome was ever conducted.

Dr. Huber also explained that glyphosate is a chelator, a small molecule that binds tightly to metal ions. In plant physiology, glyphosate's chelation disrupts a plant's uptake of essential minerals from the soil, including zinc, copper, manganese, magnesium, cobalt, and iron. Studies have shown that plants exposed to glyphosate take up much smaller amounts of these critical minerals into their tissues.[6] When we eat foods derived from these nutrient-deficient plants, we become nutrient deficient, as well.

Glyphosate also interferes with the symbiotic relationship between plant roots and soil bacteria. Surrounding the roots of a plant is a soil zone called the rhizosphere that is teeming with bacteria, fungi, and other organisms. As I will explain later in more detail, glyphosate kills the organisms living in the rhizosphere, which then interferes with a plant's nitrogen uptake, as well as the uptake of many different minerals.[7] This interference further translates into mineral deficiencies in our foods. Glyphosate also causes exposed plants to be more vulnerable to fungal diseases.[8] And fungal diseases can lead to contamination of our foods with mycotoxins produced by pathogenic fungi.

I came away from Dr. Huber's lecture convinced that I needed to learn a lot more about glyphosate.

I am a senior research scientist at the Massachusetts Institute of Technology, one of the most innovative research universities in the world. I have earned four degrees from MIT: a bachelor of science in biophysics, and a master's, an engineer's, and a PhD degree in electrical engineering and computer science. For over four decades I've worked at the intersection of human biology and computers. For my PhD thesis, I developed a computational model for the human auditory system. In my decades of research at MIT, I have published over 200 peer-reviewed scientific articles on everything from auditory modeling, conversational computer interfaces, and second language learning, to geophysics, gene structure prediction, toxicology, and human health and disease.

When I earned my PhD in 1985, I accepted a job as a research scientist at MIT, and I began a career developing multimodal dialogue systems to facilitate "natural" conversational interaction between humans and computers. Our research involved constructing interactive demonstration systems that were precursors to products like Apple's Siri and Amazon's Alexa. We also designed and developed dialogue-based computer games to help students master a second language, and we specifically focused on English-speaking students learning Chinese. I've worked to enrich people's lives using technology, to improve access to information, and to provide entertaining ways to advance language skills. Over time, I was promoted to principal research scientist and ultimately senior research scientist, the highest level on the research track at MIT.

Since 2008, I have brought my expertise in statistical analysis, computational modeling, and biology to investigate the impact of nutritional deficiencies and environmental toxicants on human health, including Alzheimer's disease, cardiovascular disorders, immune dysfunction, and neurological disorders. I have now published more than three dozen peer-reviewed scientific papers on health-related topics. I have been researching, writing about, and lecturing on glyphosate for nearly a decade. The book you hold in your hands is a culmination of that research.

As we'll explore together, there is a growing body of scientific evidence that shows that glyphosate is a major factor in several debilitating neurological, metabolic, autoimmune, reproductive, and oncological diseases. This organic chemical compound—$C_3H_8NO_5P$—is much more toxic to all life forms than we have been led to believe. Glyphosate's mechanism of toxicity is unique and diabolical. It is a slow killer, slowly robbing you of your good health over time, until you finally succumb to incapacitating or life-threatening disease. Its insidious, cumulative mechanism of toxicity, which begins with the seemingly simple substitution of glyphosate for the amino acid glycine during protein synthesis, explains the correlations we are seeing with diverse diseases that seem to have little in common.[9]

Both of my parents grew up on family farms in small towns in southern Missouri. The area is now an environmental and economic wasteland, because large agrochemical farming has forced most small farmers into bankruptcy. As a child, I visited my grandparents on their farms, gathering eggs from the chicken coop, marveling over the cows and their calves in the fields, and helping with the fruit stand where my dad's parents sold apples and peaches. When I was 13, my grandfather was discovered dead on his tractor, with a split-open bag of DDT by his side.

In the 1940s and 1950s, Americans were told that herbicides and insecticides, such as DDT, were safe. DDT is an organochloride first used by the military during World War II to control body lice, bubonic plague, malaria, and typhus.[10] While DDT was effective at preventing malaria, the environmental consequences of its use were devastating, especially as people began using it more and more, in broader and broader applications, for pest control.

I read Rachel Carson's book *Silent Spring* in 1962, shortly after it was published. A marine biologist by training, Carson condemned the chemical industry for its irresponsible disinformation campaign. She painted a grim picture of no birds singing in the spring. She called it "fable for tomorrow," a phrase that haunts me to this day. *Silent Spring* explores in detail how DDT and other chemicals were poisoning wildlife—from earthworms in the soil to juvenile salmon in the rivers and oceans. Carson's book had a profound effect on me and helped me understand my grandfather's untimely and unexpected death.

Around the same time, I also learned about the thalidomide disaster. Thalidomide, manufactured by a German pharmaceutical company, was prescribed to pregnant women to help with morning sickness and difficulty sleeping. It was aggressively marketed and advertised as safe. But thousands of children whose mothers took thalidomide during pregnancy were born with birth defects, including missing arms and legs. Studying the photographs of these deformed and unhappy children in a magazine, I realized that sometimes the products that purport to improve our lives can have major adverse effects—and that the companies that sell them cannot necessarily be trusted to tell us the whole truth about the risks their products pose.

The United States avoided this disaster, which devastated the lives of at least 10,000 children in Europe, because of a brave scientist named Frances Oldham Kelsey. Dr. Kelsey was a Canadian-born reviewer for the US Food and Drug Administration, responsible for approving or rejecting the application for a license to distribute the drug in the United States. Although she faced enormous pressure, and although thalidomide was already approved for use in Canada, Great Britain, and Germany, Dr. Kelsey rejected the application after she determined that there was insufficient evidence that it was safe to use during pregnancy.[11] At the time, I was young, optimistic, and patriotic. I remember thinking how lucky I was to live in the United States, a country that protected its citizens from such a catastrophe.

In the 1950s, in the small town in coastal Connecticut where I grew up, living treasures were everywhere: ladybugs, dragonflies, butterflies, bumblebees, grasshoppers, lightning bugs, giant beetles we called pinching bugs, toads, and dozens of chittering playful squirrels. Praying mantises were a rare delight, but fireflies could be counted on in the evening, along with bats overhead as the shadows grew. Today I live outside Boston, in a place that has a similar climate to the Connecticut town where I spent my childhood. Yet it's rare to see wildlife on our suburban street. An occasional squirrel, and one or two butterflies in the spring. No longer do we have to clean the windshield of all the dead bugs that accumulate on a summer's day.[12] Children, of course, don't realize what they're missing out on. This change appears to have happened slowly enough that almost nobody has noticed.

Yet, there's no question that something devastating *is* going on, even if it's difficult to name it precisely. The rate of species going extinct today is hundreds or even thousands of times faster than it has been during the past tens of millions of years. Environmental scientists warn that we have already entered the sixth mass extinction.[13] Human health is also suffering. Over the past few decades an alarming rise in many chronic diseases across the globe has occurred, especially in countries that adopt a Western-style diet based on industrialized agriculture. Many of these diseases have an autoimmune component. They include Alzheimer's disease, autism, celiac disease, diabetes, encephalitis, inflammatory bowel disease, and obesity.

Something terrible seems to be affecting every living thing on the planet—the insects, the animals, and the health of human beings, including children. Something hiding in plain sight. While we can't reduce all environmental and health problems to one insidious thing, I believe there *is* a common denominator. That common denominator is glyphosate. My goal, by the end of this book, is to prove to you that I am right.

My argument, as you will see, is based on connecting the dots in the peer-reviewed science. Some of the scientific arguments that I present in this book are controversial, and some conventional scientific researchers won't accept them. But this book brings together over 10 years of research that clearly shows how glyphosate is eroding both human and planetary health, resulting in a toxic legacy we are leaving future generations to contend with. This problem is too important to ignore. The goal of this book is to convince anyone who eats, anyone who has children, and anyone who cares about the health of humans and the planet that we need to look much more closely and much more carefully at the impact of glyphosate on and beyond the food supply. Both the scientific community and our regulatory establishments have failed us. It is time to shine light onto the shadows—to convince the world about glyphosate's diabolical mechanism of toxicity and give ourselves the tools we need to understand how glyphosate harms us and what we can do to protect ourselves and our families.

In chapters 1 and 2, I reveal the history of glyphosate—what it really is, how and why it was developed, and how it "works" as an herbicide. In these chapters I explore the rapidly growing body of scientific research that shows glyphosate's devastating impact on ecosystems and wildlife. In chapters 3 through 6, I move more specifically into exactly how glyphosate impacts the human body: how it damages the gut microbiome, how it substitutes for the amino acid glycine during protein synthesis, and how it disrupts the all-important and little understood roles of phosphate and sulfate. In chapters 7 through 10, I look at how this biochemistry plays out in specific conditions—liver disease, infertility, neurological disorders, and autoimmunity—that are caused in part by glyphosate's unique mechanism of toxicity. The final chapter is a call to action to safely rid the world of glyphosate, and to return to sustainable, or, even better, renewable organic agricultural practices. This last chapter also contains the best advice I can offer for how to take control of your health.

In recent years, glyphosate has gotten considerably more attention because of lawsuits that have linked it to cancer. Anyone who has read the scientific literature, even the most conventional medical doctors, understands now that glyphosate is carcinogenic, priming the body to fall prey to cancer. Throughout the book I address evidence that glyphosate causes physical damage leading to cancer, but I have chosen not to devote a chapter specifically to cancer. Why? Because cancer is the end of the line. What I want you to understand is how exposure to glyphosate sets the stage—through severe metabolic disruption—to take a person to cancer's door.

Our path forward is twofold. To state it bluntly, we need to ban glyphosate worldwide. Banning this toxic chemical is the only real way to protect what we hold dear. Until that happens, and in the absence of proper regulatory oversight, we need to protect our own health and the health of our children.

Maybe you're an environmentalist who is worried that we humans are destroying the planet, slowly poisoning the soil, waterways, plants, and animals. Maybe you're a farmer concerned about crop yields and pests. Maybe you're a science geek like me, a researcher or medical doctor or computer scientist. Or maybe you're a health care professional wanting to get to the bottom of the epidemic of disease and poor health among children and young adults you see every day in your office. Or you're a parent (or hope to be one soon) and you're desperate to figure out why so many couples fail to conceive and so many of our children are chronically ill. Whoever you are, I am glad you're here.

I wrote this book to blow the whistle on one of the most toxic chemicals of our time, to inspire you to understand the science behind how we are being slowly poisoned by glyphosate, and to spur you into action. We got into this mess because of the greed of chemical companies that put profits over people. We can get out of this mess by insisting that people and the planet matter more. The science will lead us where we need to go. I invite you to journey with me.

— CHAPTER 1 —

Evidence of Harm

Future historians may well look back upon our time and write, not about how many pounds of pesticide we did or didn't apply, but by how willing we are to sacrifice our children and future generations for this massive genetic engineering experiment that is based on flawed science and failed promises just to benefit the bottom line of a commercial enterprise.

—DON HUBER, PhD

Glyphosate-based herbicides are used to control a wide variety of weeds that grow among food crops, residential lawns, gardens, public parks, roadsides, conservation lands, wildlife areas, rangelands, forests, waterways, and more.[1] These herbicides come under a long list of names, including Roundup, Roundup Ultra, Roundup Pro, AquaMaster, Aqua Neat, Polado, Accord, Rodeo, Touchdown, Backdraft, Expedite, EZ-Ject, Glyfos, Laredo, Buccaneer Plus, and Wrangler, among others. These products contain a variety of chemicals. But in all of them, glyphosate is the primary ingredient, making up 36–48 percent of the product.

Although it's used as a weed killer today, glyphosate was first patented by the Stauffer Chemical Company in 1961 as a chelating agent to strip mineral deposits off pipes and boilers in commercial hot-water systems.[2] Then in 1968, Monsanto patented glyphosate for a totally different application, as an herbicide for use in agriculture.[3] It was then patented a third time (again by Monsanto) in the early 2000s, this time as an oral antibiotic.[4] All of these applications, as a chelator, an herbicide, and an antibiotic, play a role in the chemical's unique and diabolical impact on human health.

With the legal right to its exclusive use in agriculture, Monsanto began selling a glyphosate-based herbicide, Roundup, in the mid-1970s. Roundup was marketed as being considerably less toxic than other common herbicides, such as dicamba and 2,4-D, less persistent in the soil, and safer for humans, animals, and the environment. By reducing the need for tillage, it was described as "the perfect environmental solution at the perfect time . . . one of, if not the safest, herbicides in history."[5]

As a nonselective herbicide, it would kill almost any plant it came in contact with. Farmers had to be cautious in its application. In the 1980s, Monsanto began investing in a brilliant new biotechnology research and development program. The company's genetic engineers spliced microbial genes into common crops to make them resistant to Roundup. We generally use the term *genetically modified organisms* (GMOs) to describe these types of crops. These "Roundup Ready" GMO crops transformed the herbicide from a home run to a grand slam. By the time glyphosate-resistant crops came on the market in 1996, they were heralded as a major advance in agriculture and biotechnology—a promise to end hunger and feed the world. And instead of using glyphosate carefully and selectively, farmers could now douse entire fields with it and only the weeds, not the Roundup Ready crops, would die.

Sales of Roundup Ready seed, beginning with soybeans and corn, and followed by canola, sugar beets, cotton, and alfalfa, skyrocketed. So did sales of Roundup. By 2014, glyphosate use in the United States had risen to 125 million kilograms (about 138,000 tons), up 15-fold from the 1970s.[6] Today nearly 150,000 tons of glyphosate are sprayed onto American crops every year. That's the equivalent of one pound of glyphosate per year for every person in the United States.[7]

From an industry standpoint, there was only one big monkey wrench in the weed-killing glory of glyphosate: Living beings want to survive. Organisms that can withstand poisons will flourish and multiply. While glyphosate would kill weeds indiscriminately, certain weeds were less sensitive—or became so over time. These "superweeds" begin to dominate an ecosystem, crowding out crops, requiring more and more glyphosate, or different herbicides entirely, to kill them off. The solution? Use more glyphosate! Since 1974, about 8.6 billion kilograms—some 19 billion pounds—of glyphosate have been applied worldwide.[8]

GMO technology is a powerful tool that allows genes from other species (usually microbial genes) to be introduced into the genome of a plant. Such novel genes can supply the plant with beneficial advantages—frost resistance, for example, or drought tolerance. To produce glyphosate resistance, a gene from a bacterium with a version of EPSP synthase insensitive to glyphosate is inserted into the crop's genome. The amount of glyphosate used on GMO crops has increased dramatically over the past two decades. Given the hand-in-glove relationship between glyphosate and genetically modified crops, you might think it's easy to avoid glyphosate by avoiding GM foods. In fact, the highest levels of glyphosate have consistently been found in *non-GMO* foods derived from wheat, oats, and legumes. Why? Because these crops are commonly sprayed with glyphosate as a desiccant right before harvest, causing the plant to drop its leaves and dry out so it is easier to harvest. Glyphosate forces the crop to go to seed as it's dying, which synchronizes seed production and increases yield. Many other crops are desiccated with glyphosate, including barley and rye, and oily crops such as canola, safflower, sunflower, linseed and flax, that are used for vegetable oil production. At this point in the plant's life cycle, it is not a problem if glyphosate kills it. In fact, that is the intent.

The astute shopper should be aware of the difference between the label "non-GMO" and "certified organic." By law, glyphosate cannot be used on certified organic crops. Even if a food is certified organic, however, it doesn't necessarily mean the food is glyphosate-free. Though organically grown foods usually test at much lower levels for glyphosate than conventionally grown food, it's nearly impossible to avoid glyphosate in the soil, in animal manure, in rainwater, and in wind drift.

Glyphosate-based herbicides have become so pervasive that even food from farms nowhere near where glyphosate is sprayed can be contaminated. When the FDA tested 28 samples of honey in 2017, 100 percent contained glyphosate.[9] Two years later, government scientists in Canada found the herbicide in 197 out of 200 honey samples they tested.[10] Bees don't understand property boundaries. They forage freely, picking up bits of herbicides and insecticides as they fly from flower to flower, bringing it back to their hives.

Glyphosate has been showing up in human urine since 1993, well before genetically engineered Roundup Ready crops were in widespread use.[11] But while only 12 percent of Americans tested positive for glyphosate exposure from 1993 to 1996, at least 70 percent of Americans test positive today.[12] As high as that number sounds, it's likely an underestimate. Of the several dozen patients one doctor in southern Oregon tested for glyphosate—most of whom were self-described health nuts who eat organic food—100 percent, including the doctor himself, came back positive.[13] People who eat a predominantly organic diet have significantly less glyphosate in their urine than people who consume mostly conventional foods, and people who are healthy have significantly lower levels of glyphosate in their urine than those who are chronically ill.[14] Still, glyphosate is nearly impossible to completely avoid.

We are all being exposed to glyphosate—primarily from the food we eat, but also from the water we drink and bathe and swim in; the lawns we walk on; the parks children play at; and probably even the medicines and nutraceuticals we take. Glyphosate is also in the air—a particular risk to people who live in farming communities—and it has been shown to be a contributing factor in rising rates of lung damage, asthma, and allergies.[15] When a chemical is so pervasive, so ubiquitous, so nearly impossible for even the most diligent person to avoid, it is especially incumbent on regulatory agencies and elected officials to ask tough questions, conduct rigorous investigations and hearings, and put the health and safety of its populace first. But in the case of glyphosate, this hasn't happened. It's an abdication of responsibility and a disgrace to democracy.

Junk Science

In 2020, a team of Italian medical doctors specializing in cardiovascular disease analyzed emails and memos from Monsanto, finding that the company had a "coordinated strategy to manipulate the debate about the safety of glyphosate to the company's advantage." Similar to the junk science produced and promoted by the tobacco industry, this manipulation "seriously jeopardizes the credibility of . . . scientific study in the modern era," the Italians concluded.[16]

In 2019, a team of California-based scientists tried to parse out whether genetically modified foods per se were damaging, or whether it was the glyphosate that genetically modified foods are frequently sprayed with that poses the real threat.[17] Their experiments revealed that when fruit flies were fed genetically modified foods that were treated with glyphosate, they died sooner and in greater numbers; but when they were fed genetically modified foods that weren't treated with glyphosate, they didn't. The problem is that industry-funded studies to evaluate the safety of genetically modified organisms are often conducted on animals fed genetically modified foods that are not treated with glyphosate.[18] You read this correctly: They test foods that are not treated with glyphosate even though the point of genetic modification is to allow crop plants to withstand glyphosate. In practice, genetically modified crops are almost always exposed to glyphosate. In order to prove this food is safe, we have to test crops that have been exposed to glyphosate.

Well-designed toxicology studies include a control, a group of subjects identical to the treatment group except for the toxic exposure. Yet even control groups of animals purportedly not exposed to glyphosate often have detectable levels of glyphosate in their bodies.[19] This may be due to glyphosate contamination of animal feed, water, bedding, or all three. This means there is often no real glyphosate-free control group to compare against glyphosate-exposed subjects. We need to keep this in mind when evaluating papers that compare a control group to a treatment group of glyphosate-exposed animals.

Acceptable Amounts?

From the 1920s until it was finally phased out in 1996 as part of the Clean Air Act, lead was a key additive to gasoline in the United States.[20] Houses were painted with lead paint and water pipes were made from lead. It's a remarkably useful metal, but it's also extremely harmful to human health. When my four sons were young, lead wasn't on any doctor's radar. In the 1980s doctors were told that children could safely have 20 micrograms per deciliter of lead in their bloodstream. Now we know that even just 5 micrograms of lead in the blood is dangerous and potentially brain damaging. The current scientific consensus is that no amount of lead is safe in the bloodstreams of children.

The Environmental Protection Agency's job is to evaluate and regulate pesticides to make sure they are safe for both human health and the environment. (Insecticides, fungicides, and herbicides all fall under the umbrella term *pesticide*.) The Food and Drug Administration's job is to make sure chemical herbicide food residues don't exceed the EPA's safety limits. Since first registered with the EPA, glyphosate has been reviewed for safety once every 15 years.[21] In 2016, the EPA declared that the Acceptable Daily Intake of glyphosate is 1.75 micrograms per kilogram of body weight per day.[22]

(The ADI adopted by the European Union is much lower at only 0.5 micrograms per kilogram of body weight per day.[23]) This means that, according to the EPA, a 150-pound person (68 kilograms) can "safely" eat 120 micrograms of glyphosate a day. In January 2020, in response to industry pressure, the EPA declared that "there are no risks of concern to human health when glyphosate is used in accordance with its current label," and that glyphosate is "unlikely to be a human carcinogen."[24] But science shows that the EPA is wrong. As I explain throughout this book, a growing body of scientific literature strongly suggests that *no amount* of glyphosate is safe.

In 2018, a team of scientists from Italy, Denmark, and the United States exposed albino rats to glyphosate and Roundup at the rodent-equivalent doses considered safe for humans, and then tested their microbiomes. The microbiomes of both the mother rats and their offspring were disrupted, the pups faring worse than the dams. The exposed rats showed fewer beneficial bacteria, such as *Lactobacillus* and *Bifidobacteria*, and more pathogenic ones—*Prevotella*, in particular, which is often associated with infections. *Lactobacillus*, normally the first species to colonize the gut, aid in metabolizing milk. Disrupting *Lactobacillus* allows pathogenic bacteria to gain traction. The glyphosate-exposed rats suffered many other negative health effects, including hormone problems and reproductive damage. Female rats also had abnormally high levels of testosterone.[25]

In another well-designed experiment, Italian biologists exposed human cells to glyphosate at varying concentrations, most at amounts below the EPA's ADI. The glyphosate-exposed cells showed abnormally high chromosomal aberrations at all levels except the very lowest.[26] Experiments on eels and fish have similarly shown DNA damage after exposure to glyphosate-based herbicides at very low levels.[27] Likewise, researchers in Spain have found that glyphosate crosses the blood-brain barrier and overstimulates neurotransmitter receptors in various regions of the brain, causing neuronal excitotoxicity, a pathological process that damages and kills neurons. When the researchers investigated different exposure levels, they uncovered a dose-response relationship between glyphosate and the major neurotransmitters in different regions of the brain.[28]

Even at exposure levels measured in *parts per trillion*, glyphosate stimulates hormone-sensitive breast cancer cells to grow unchecked.[29] A part per trillion is one drop of water in an Olympic-sized swimming pool, an amount that is so tiny it's hard for our brains to conceptualize. Glyphosate primes mammary cells to become more sensitive to well-established carcinogens, increasing the risk that these toxicants will induce cancer.[30] Researchers have also found that when human lung cells are exposed to Roundup at concentrations ranging from 50 to 125 micrograms per milliliter—lower than the typical occupational exposure level of agricultural workers—it causes DNA damage and programmed cell death.[31] DNA damage over time can lead to cancer.

Scientists who study hormones in humans and other animals have found that endocrine-disrupting chemicals often have more dramatic effects at lower doses than at higher doses, defying the adage that "the dose makes the poison."[32] In fact, counterintuitively, glyphosate-associated birth defects are often *more frequent* with a lower-dose exposure during pregnancy. After exposing pregnant rats to relatively high doses of a glyphosate formulation, a team of scientists from across the United States and Korea found an inverse relationship between the exposure level and the presence of malformations. The highest percentages of pups with the most health problems, including a severe generalized whole-body tissue swelling related to liver failure, kidney failure, and heart failure (called anasarca), were those exposed to the *lowest* amounts of glyphosate. If you're an animal lover, the details from this experiment are difficult to read: Half the pregnant rats died from the toxic exposure.[33]

Killing Us Slowly

Rats and other rodents make up 95 percent of all lab animals used in biomedical research. There are many reasons why they're a good choice: They're small, easy to maintain in cages, and relatively inexpensive to care for. Rats and mice also have a lot in common with humans in terms of genetics and disease processes. Their one- to two-year life spans mean that multiple generations can be studied over a short time period. Sophisticated biotechnology also enables creative design of rodents' genetic profile, including the insertion of human genes into their genome so as to better understand effects on humans. The genome of rats is well characterized, and many species have been inbred in the laboratory over decades to create specific strains with little genetic variability and predictable metabolic features. These "bioidentical" rodents are especially helpful in identifying and studying environmental toxicants.

In all the hundreds of scientific studies I have read, one stands out. French researchers, led by Gilles-Éric Séralini, a professor of molecular biology at the University of Caen Normandy, showed that chronic exposure to low doses of Roundup leads to significant harm in rats. In a study that essentially repeated a study that had been conducted by Monsanto, Séralini's team fed rats Roundup-treated genetically modified corn, but instead of stopping at three months, as Monsanto did, Séralini's team continued the experiment for two years.[34] This time frame is significant, because the agrichemical industry has devised an expedient rule that three months is sufficient time to demonstrate toxicity. While there were no obvious differences between the control group and the experimental group at three months, the female rats eventually developed massive mammary tumors, male rats experienced damage to their liver and kidneys, and both males and females experienced reproductive problems and early death. The first two male rats that died in the experimental group died a year earlier than the first male in the control group. *Three times* as many females in the group that ate Roundup-exposed corn died by the end of the two-year experiment as in the control female group.

Following the 2012 publication of this research in *Food and Chemical Toxicology*, Monsanto pressured the journal for a retraction, mainly arguing that the number of rats in the study was too small.[35] One former Monsanto scientist, Richard Goodman, used his position on the journal's editorial board to put pressure on the journal's editor, a former Monsanto employee, to discredit the research. The journal's retraction decision is almost comical: "Unequivocally, the Editor-in-Chief found no evidence of fraud or intentional misrepresentation of the data," the journal editors explained. But it also said: "Ultimately, the results presented (while not incorrect) are inconclusive."[36]

Scientific results are always open to further scrutiny. To retract a paper because its results are "inconclusive" is highly irregular. That Monsanto succeeded in its aggressive campaign to censor the research, and the journal issued a retraction in November 2013, demonstrates a disturbing level of industry control over scientific and public debate. In 2014, Séralini's paper was republished in *Environmental Sciences Europe*.[37]

Séralini's team published another paper in 2018 that analyzed toxicity of 14 different glyphosate-based formulations.[38] Using mass spectroscopy, the team identified a family of petroleum-based oxidized molecules, including polyethoxylated tallow amine (POEA), as primary additives that increase a plant cell's ability to take up glyphosate. Mass spectroscopy also detected the heavy metals arsenic, chromium, cobalt, lead, and nickel. The glyphosate formulations acted as endocrine disruptors at levels below the plant cell's cytotoxic threshold, whereas glyphosate alone did not. These findings are especially disturbing because the initial studies conducted to evaluate glyphosate's toxicity, as well as the ongoing reevaluations that have led to its continued approval, only evaluate glyphosate in isolation. Scientific findings about safety that may be technically accurate in a petri dish may not translate into the real-world use of this chemical. Why? Because in practice, glyphosate is never used in isolation.

Séralini and his team concluded that glyphosate formulations are much more acutely toxic than glyphosate alone. Acute toxicity refers to the negative effects of a substance after a single exposure or after multiple, but short-lived, exposures. I believe these scientists are correct in that we should be concerned about glyphosate and its adjuvants causing acute toxicity. At the same time, as we will see, it is likely that the slow kill effect observed in their long-term study is due to glyphosate itself, not to the other ingredients.

The industry has led us to believe that ingested glyphosate is swiftly removed from the body via the urine and the feces, but many independent studies have shown otherwise. Glyphosate has been detected in the kidney, liver, lungs, spleen, muscles, and intestines of dairy cows, with the highest levels showing up in the lungs.[39] Residues have been found in the liver, spleen, lung, intestine, heart, muscles, and kidneys of broiler chickens.[40] And glyphosate has also been detected in multiple organs of deformed piglets—primarily in their hearts and lungs, but also in lower amounts in their muscles, livers, kidneys, brains, and intestines. These glyphosate-exposed piglets were born with congenital malformation of their ears, brains, and legs. One had only a single large eye, another an elephant-sized tongue, a third piglet was missing a snout. Yet another was a female piglet born with testes.[41]

A Powerful Chelator

As I mention in the introduction, chelators are small molecules that bind tightly to metal ions. *Chele* in Greek means claw-shaped, so think of chelators as molecules with claws that "hook" metals. In biology, chelation can render the metal ions inaccessible to living cells, so doctors will use a powerful chelator such as EDTA (ethylenediamine tetra-acetic acid), which can bind with many minerals and metals including aluminum, zinc, magnesium, and calcium, to treat life-threatening hypercalcemia and other toxic metal exposures. Deferoxamine binds with iron and aluminum, so it is sometimes used to remove iron that builds up in the blood from repeated transfusions. Since the 1950s, chelators have also been used to treat lead exposure.

Though chelators can have beneficial applications, they can also be dangerous if they bind with minerals that play an essential role in living cells. Glyphosate is a powerful chelator. It binds tightly to metals, making them less available as catalysts for multiple enzymes that depend on them for proper function. (A catalyst is something that increases the rate of the reaction without itself being modified; many minerals serve this role in enzymatic reactions.) Glyphosate is particularly effective at binding to a class of minerals called +2 cations, including zinc, copper, manganese, magnesium, cobalt, and iron. Since plants exposed to glyphosate take up smaller amounts of these critical minerals into their tissues, foods derived from these plants will be mineral deficient.[42] We humans need only tiny amounts of these minerals in our diet, but deficiencies can result in serious adverse health consequences.

Let me give you an example. Many pancreatic digestive enzymes depend on zinc. Zinc deficiency can disrupt the immune system, cause unexplained weight loss and brain fog, and lead to diarrhea, loss of appetite, skin sores that won't heal, and a loss of sense of smell and taste. Zinc deficiency is more prevalent in animals chronically exposed to glyphosate. In a German study, piglets fed a zinc-deficient diet showed serious digestive problems early on, even before clinical symptoms of zinc deficiency appeared.[43]

Cobalt and manganese are also affected by glyphosate. In a study on dairy cows in Denmark, these minerals were consistently found to be at levels far below the *minimum* expected range in cows from eight different farms.[44] Cobalamin, which you might know of as vitamin B_{12}, depends on cobalt as a catalyst. It's an important enzyme in the body. It regulates many metabolic pathways, and it is vital for the normal formation of red blood cells and healthy nerve and brain tissue. Only a few enzymes depend on cobalamin, but each is essential for cellular function, a subject I'll return to in chapter 3 and chapter 9.

Simultaneously Deficient and Toxic

Anthony Samsel is a unique scientist: brilliant, passionate, eccentric, and stubbornly self-sufficient. He had an illustrious career as a chemist in an Arthur D. Little think tank, and he holds many chemical patents. Now retired, Samsel lives on several acres of land in New Hampshire where he grows organic crops, providing most of the food for his family from his own small farm. Samsel also still works as a toxicology consultant. His expertise includes assessing the nutritional content of food crops exposed to toxicants and studying contamination in foods, nutritional supplements, and medicines. He has been collecting important data on glyphosate levels in teeth, fingernails, bile acids, and digestive enzymes.

Samsel and I met in 2012. Since then we have collaborated on a series of six peer-reviewed scientific articles. Together we discovered that glyphosate not only makes beneficial minerals toxic, but that it also transports and delivers known toxic metals, such as aluminum and arsenic, to acidic areas of the body where it then releases the toxic cargo. In our research we found that glyphosate disrupts manganese, making it simultaneously deficient and toxic.[45] Other scientists have confirmed that by escorting arsenic to the kidneys and unloading it in the acidic environment of the renal tubules, glyphosate is likely contributing to an epidemic in kidney failure in Sri Lanka.[46] Kidney failure at a young age is also now a major problem among agricultural workers in the sugar cane fields in Central America, and glyphosate is likely a major contributor here as well.[47]

You have just learned that EDTA is a strong metal chelator. Glyphosate binds with aluminum one million times more readily than EDTA does.[48] Two glyphosate molecules wrap around an aluminum atom, hiding its charge and producing an uncharged small molecule that easily crosses barriers.[49] This glyphosate binding allows aluminum to be carried past the gut barrier and into the brainstem nuclei, where an acidic environment prompts the glyphosate to release it.[50] The last place we want aluminum is inside the brain.[51]

In December 2015, Anthony called me up to tell me of his excitement in his realization of the chilling idea that glyphosate might be disrupting protein synthesis through its role as an amino acid analogue of the coding amino acid glycine. I was skeptical at first, but once I started investigating this idea, it became clear to me that it could be the mechanism by which glyphosate is a diabolical and insidious disruptor of systemic metabolism in all living species. I'll be walking you through this concept, which is a central theme of this book, later on.

Killing Us Quickly

Farmers have been dying by suicide in record numbers in both the United States and around the world.[52] These agricultural workers have an easy way to harm themselves: by drinking the poisons they use on their crops. There is evidence that death by glyphosate intoxication has become a serious problem in Argentina, Brazil, Canada, and the United States.[53] Symptoms of acute glyphosate poisoning include intestinal pain, vomiting, fluid buildup in the lungs, pneumonia, loss of consciousness, difficulty breathing, loss of muscle control, convulsions, destruction of red blood cells, and death.[54]

Of course, these suicides and suicide attempts involve drinking a glyphosate-based *formulation*, which has other ingredients besides glyphosate. These ingredients include salt buffers and surfactants such as POEA, which are added to act as adjuvants to make glyphosate more soluble and better able to enter plant cell walls. As we saw from Séralini's research, these other ingredients may be acutely toxic in and of themselves. In fact, in some cases, they may be up to a thousand times more acutely toxic than glyphosate.[55]

While it makes you tremendously ill, ingesting glyphosate formulations is not always lethal. Research from Taiwan showed that over 70 percent of people with acute exposure survived.[56] Metabolic acidosis, an abnormal chest X-ray, tachycardia, and kidney stress were associated with an increased risk of dying, as were dangerously elevated potassium levels and breathing problems severe enough to require intubation. From this research, conducted in 2008, it looks as though damage to the lungs, kidneys, and heart are key factors leading to death.

Other analyses have shown survival rates as low as 54 percent and as high as 92 percent.[57] In 2011, when Taiwanese researchers studied 131 patients, they found that every one of them became very ill. Altered consciousness, difficulty breathing, irregular heartbeat rhythm, and shock were more frequent in people who died. One of the reasons drinking a glyphosate-based herbicide can be deadly is that it suppresses cardiac conduction and contractility and essentially destroys the heart. A South Korean study conducted in

2014 analyzed data from over 150 patients who attempted suicide by this method.[58] Twelve percent died. As was the case in Taiwan, these deaths were often preceded by severe heart arrhythmia.

Cancer Concerns

In April 2015, after looking closely at all the existing literature on glyphosate, the International Agency for Research on Cancer, which is part of the World Health Organization, declared glyphosate a probable carcinogen.[59] This designation met with considerable pushback from Monsanto. But the IARC stood firm. People who developed cancer, specifically non-Hodgkin's lymphoma, after using Roundup became emboldened to seek compensation from Monsanto.

Dewayne Johnson was a groundskeeper at a California school in 2014 when he was diagnosed at the age of 42 with non-Hodgkin's lymphoma. He had been using Roundup frequently as a part of his duties. In a historic lawsuit, this father of three young sons sued Monsanto, now owned by the German multinational Bayer AG, alleging the company had caused his life-threatening cancer by covering up the known risks of Roundup for decades. In a landmark verdict, Johnson was awarded $289 million by a jury (later reduced by a judge to $78 million), emboldening other victims and paving the way for future lawsuits.[60] "Monsanto does not particularly care whether its product is in fact giving people cancer," Judge Vince Chhabria wrote in a court document filed on March 7, 2019, "focusing instead on manipulating public opinion and undermining anyone who raises genuine and legitimate concerns about the issue."[61]

Cancer can take a long time to develop. As I mentioned previously, excessive DNA damage is a precursor to cancer. Scientists test for a chemical's potential as a carcinogen by exposing cells to the chemical and looking for damage to the chromosomes. When human liver cells are exposed to low doses of glyphosate, similar to what we're exposed to in the environment, DNA damage ensues.[62] Human white blood cells exposed to glyphosate also suffer from DNA damage, including to a gene that suppresses cancer tumors (the p53 promoter gene).[63] Destroying the protective benefit of the p53 promoter gene paves the way for invasive leukemias and lymphomas.[64] We also see in population science what we see in the laboratory: Researchers have found DNA damage in agricultural workers exposed to glyphosate in Ecuador and Colombia.[65]

While we have been told for years by industry apologists that Roundup is safe, now even mainstream news outlets know we've been duped.[66] In June 2020, Bayer announced a settlement to pay over $10 billion to settle 100,000 lawsuits with plaintiffs who claimed they got cancer from using its products.[67] One of the largest settlements in the history of US civil litigation to date, it isn't nearly enough. No amount of money in the world can give people back their health. Bayer has admitted no wrongdoing or liability.

The world needs to understand that cancer is only one of the health hazards of glyphosate exposure.[68] Glyphosate pollutes the soil and the plant life that depends on it. It taints our food and our water. The negative health consequences of glyphosate can be seen across multiple generations. It's biopersistent and, in the United States especially, almost unavoidable. Supposedly safe limits for humans are based on outdated science.

The healthiest countries are those where people are able to live out their natural lives without getting sick and dying prematurely. The life expectancy rate in the United States lags sorely behind its industrial counterparts. While all other industrialized countries are seeing life expectancy rates rise, ours declined from 2014 to 2017.[69] For one of the wealthiest countries in the world, we have among the lowest life expectancy rates. Even the death rate among young people in America is going in the wrong direction.[70]

On nearly every health indicator the United States ranks last or next-to-last among industrialized countries. We also use more glyphosate per capita than any other industrialized nation. Of course, correlation does not equal causation, as many working in the chemical industry, and those who turn a blind eye to our current health crises, are quick to point out. But in the case of glyphosate, as I'll show you throughout this book, there's a more apt expression: Where there's smoke, there's fire.

— CHAPTER 2 —

Failing Ecosystems

If our extinction proceeds slowly enough to allow a moment of horrified realization, the doers of the deed will likely be quite taken aback on realizing that they have actually destroyed the world.

—ELIEZER YUDKOWSKY, research fellow,
Machine Intelligence Research Institute

The words of ecologist William Ophuls haunt me: "A fortunate minority gains luxuries and freedoms galore," he wrote in an essay titled "Apologies to the Grandchildren," "but only by slaughtering, poisoning, and exhausting creation. So we bequeath you a ruined planet that dooms you to a hardscrabble existence, or perhaps none at all."[1] I have four sons and eleven grandchildren. As I write these words, the youngest is just a few days old. There are about 7.8 billion people on the planet now. I do not want to hand a scorched Earth to our children and grandchildren.

There are 550 gigatons of carbon in the form of biomass on Earth.[2] Humans, despite our exploding population, account for only 0.01 percent of it. If humans were all together on a giant seesaw with every person on Earth on one side and all the bacteria on Earth on the other, we humans would shoot skyward, our legs dangling in the air high above the ground. Bacteria vastly outweigh and outnumber us. But despite accounting for so little organic matter, humans have an outsized and particularly destructive impact.

The loss of clean water, healthy soil, and unpolluted air poses a dire threat to humanity and to many wild species. Populations of mammals, birds, fish, reptiles, and amphibians have already shrunk by 60 percent in just four decades. Many species will become extinct within the next 10 years. This loss of biodiversity will have disastrous consequences. If we maintain our current practices, we run a real risk of polluting ourselves—and much of life as we know it on our planet—out of existence.[3]

As I mentioned in chapter 1, over the past two decades the use of glyphosate on core crops has increased dramatically. The introduction of the genetically modified crops in the late 1990s was a boon to the industry. Roundup Ready crops, including corn, soy, canola, sugar beets, alfalfa, cotton, and tobacco, are genetically modified by the insertion of a microbial gene that produces a version of the enzyme EPSP (5-enolpyruvylshikimate-3-phosphate) synthase that is resistant to glyphosate's effects. This is the enzyme in the shikimate pathway that glyphosate disrupts.

As you know, many non-GMO crops are also commonly exposed to glyphosate. These crops are commonly sprayed right before harvest, with glyphosate acting as a ripener or a drying agent. Particularly in northern regions such as Canada where the growing season is short, the goal is to force the crop to ripen before frost sets in. As I mentioned in the last chapter, this practice is also increasing in part because farmers have learned that synchronizing plants to go to seed at the same time improves yield. Crops commonly treated right before harvest include wheat, oats, barley, sugar cane, sunflower seeds, and legumes such as chickpeas, lentils, and soybeans.

Like the United States and Canada, Argentina has seen a rapid rise in glyphosate use over the past two decades, along with the widespread adoption of genetically modified glyphosate-resistant soy crops, mainly for export. Introduced in the 1980s, glyphosate use quadrupled in Argentina between 1996 and 2012. Some of the highest glyphosate residues have been accumulating year by year in the Pampas region, a center for soy production.[4]

Monsanto contends that glyphosate disappears quickly from the environment after it is applied. The company claims that glyphosate is largely metabolized by soil bacteria within a couple of weeks. Scientific research paints a different picture. Of five herbicides used in Finnish sugar beet fields, glyphosate was found to be the second most persistent, still present well into the spring following a fall application.[5]

Scientists have also found that, in places where glyphosate is heavily used, the total amount of glyphosate used over time correlates more strongly with the amount of glyphosate detected in the soil than the amount of glyphosate used in the most recent applications. This was found to be true in the Pampas region of Argentina where Roundup Ready soybeans are grown. Practically speaking, this means that the amount of glyphosate a farmer applies each year likely exceeds the degradation rate. With every five applications, there is an estimated increase of one milligram of glyphosate per kilogram of soil.[6]

How quickly does glyphosate actually degrade in natural soils? Slowly. In one experiment, radiolabeled glyphosate was added to undisturbed sand and clay. Scientists then analyzed samples, taken weekly for more than two years. They found that after 748 days, 59 percent of the glyphosate was still present.[7]

No one denies that there is an environmental crisis unfolding today. Different species from multiple phyla, from fungi to insects to amphibians to birds, are experiencing rapid and alarming population declines. Of course, not all environmental destruction can be attributed to glyphosate. But it is a major factor in biodiversity loss, including the dramatic collapse of monarch butterflies and honeybees, and declining overall health in plants, animals, other living organisms, and entire ecosystems. It is inducing an epidemic of pathogenic fungal infection in plants, animals, and humans. There is even evidence that glyphosate is contributing to health problems in pets.

In 2017 a team of French and English scientists found disruption in 82 proteins present in the fungus *Aspergillus nidulans* following glyphosate exposure in amounts far below what is recommended for agricultural use. They discovered that the majority of the disrupted proteins were part of the detoxification and stress response. Others were linked to protein synthesis, amino acid metabolism, and the citric acid cycle.[8] This research shows that glyphosate disrupts the cells of even the most primitive life forms in complex ways, causing them to abnormally increase production of some proteins and decrease production of others.

Fungus among Us

Although people often associate fungi and mushrooms with pathogenicity, many fungi are actually beneficial to humans, the soil, and entire ecosystems. In fact, we cannot survive without them. Mycorrhizal fungi live in the rhizosphere around plant roots, benefiting the plant and the soil ecosystem in diverse ways. With their vast networks of filamentous mycelium, mycorrhizal fungi facilitate the uptake of nutrients and minerals and increase plants' resistance to drought, salinity, insects, and toxic exposures.[9] The mycelium even forms a communication network among the plants to inform one another of threat.[10]

Mycorrhizae are part of a complicated, magnificently organized underground ecosystem of living and dead matter. Soil contains trillions of microorganisms, including fungi, bacteria, nematodes, and protozoa, as well as larger organisms such as earthworms, ants, underground insects, and burrowing animals. A healthy soil ecosystem is a biodiverse ecosystem.

Glyphosate, however, disrupts the balance. It clears the way for pathogenic fungal overgrowth, predominantly from mycotoxins produced by *Fusarium* and *Aspergillus* strains of fungi. Pathogenic species of *Fusarium* cause crown rot, root rot, and head blight in cereal crops. In Australia, *Fusarium* species are estimated to account for nearly 8 million dollars of lost revenue from reduced yield due to crown rot.[11] A study based in western Canada showed that glyphosate use was one of the biggest factors in the proliferation of pathogenic fungi.[12] The fungi produce excessive amounts of oxalic acid, and its accumulation in plant tissue leads to a drop in the plant's pH. This is a boon for fungi, which have many enzymes with optimal activity below pH 5. But the metabolites released by these fungal enzymes cause the plant to wilt.[13]

Pathogenic fungi also threaten bats, amphibians, and reptiles, as well as human health.[14] One of my nieces recently developed a *Candida* infection. She had a thick white coating on her tongue, a cottony feeling in her mouth, and difficulty swallowing. Fungus commonly causes toenails to become painfully itchy, brown, and cracked. In Central and South America, people who work outdoors in rural areas sometimes get a skin and lung disease called paracoccidioidomycosis, which causes lesions in the mouth and throat associated with swollen lymph nodes, fever, and weight loss. Overall, fungal infections of the skin, nails, and hair affect more than a billion people worldwide.[15]

More than 150 million people have fungal disease that is severe enough to be life threatening. The number of people who die of fungal disease each year is 1.7 million. Global mortality associated with fungal disease is higher than that of tuberculosis and more than three times higher than malaria.[16] Some common fungal diseases that can become serious include aspergillosis, *Pneumocystis* pneumonia, vaginal candidiasis, thrush (*Candida* in the mouth), and invasive candidiasis, where *Candida* infects multiple body organs.[17]

Glyphosate in the soil and water is likely contributing to increases in fungal infections. Many pathogenic fungi use glyphosate both as a nutrient and an energy source. The fact that some fungi thrive in the presence of glyphosate has led scientists to make a case for using certain fungal species, such as *Aspergillus oryzae*, to help *clear* glyphosate from the soil.[18] There's another connection between the rise of pathogenic fungi and the overuse of glyphosate. Recall that I noted in chapter 1 that glyphosate is patented as an antibiotic. Many bacterial species in the human gut are sensitive to glyphosate, and the loss of bacteria caused from this antibiotic gives fungi such as *Candida* an opportunity to expand disproportionately.

When scientists in Argentina compared fungi that thrived in soils exposed to glyphosate with fungi that thrived in control soils without glyphosate exposure, they found that *Candida krusei*, a wild species of yeast, were dominant in the glyphosate-exposed soils.[19] This fungus's ability to degrade glyphosate gave it a distinct advantage in the exposed soils. *Candida krusei* is now a multidrug-resistant fungus that infects humans, and it has become a major threat for patients suffering from leukemia and lymphoma, due to their weakened immune systems.[20]

You may have noticed an intriguing paradox in all this research: Pathogenic fungi that can metabolize glyphosate not only have a distinct advantage over other species in the human host but also protect the host from glyphosate toxicity by clearing the glyphosate. While there may be a future role for fungi in helping us heal glyphosate-damaged soil, the overgrowth of fungi leading to serious and potential fatal infections is cause for concern.

On November 13, 2019, the Centers for Disease Control and Prevention (CDC) released a report warning about the emergence of antibiotic resistant fungi, singling out a new species, *Candida auris*, which first showed up in Japan 10 years earlier.[21] It is now circulating among hospitals on multiple continents and is resistant to all known antifungal agents. As of November 25, 2020, there were a total of 1,364 confirmed clinical cases of *Candida auris* in the United States with the highest numbers in Illinois, New York, New

Jersey, and Florida.[22] It is extremely virulent with a 30 to 60 percent fatality rate.[23] Anyone can succumb to a pathogenic fungal infection, but people with compromised immune systems are especially susceptible. Glyphosate, as we'll see in chapter 10, compromises the immune system. Places where fungal resistance is most problematic—North America, Europe, Australia, Brazil, and India—are also places where glyphosate usage is most prevalent. We can anticipate a future fungal disease pandemic if glyphosate usage continues to escalate as it has in the past.

Poisoning Our Waters

Scientists have detected glyphosate in soils and sediment, ditches and drains, precipitation, rivers, and streams in multiple locations across the United States.[24] Glyphosate pollution was found in 84 percent of 68 water samples taken from Canada's Saint Lawrence River system.[25] And a study based on samples taken from the intensive agricultural greenbelt zone around the city of Córdoba, Argentina, found high occurrences of glyphosate in water, sediment, and suspended particulate matter.[26] Glyphosate in waterways causes harm in a variety of ways.

Lake Okeechobee in south central Florida is surrounded by sugarcane fields that are sprayed with glyphosate right before harvest, likely contributing to cyanobacteria blooms in the lake. Cyanobacteria, also known as blue-green algae, grow in fresh, brackish, and marine waters and thrive when the temperatures are high and there is a lot of phosphate in the water. Phosphate levels in waterways have been steadily rising across the United States over the past decade.[27] The percentage of streams in the United States uncontaminated by phosphate decreased from 24.5 percent in 2004 to 10.4 percent five years later to only 1.6 percent in 2014. Streams that were once relatively free of phosphate are now full of it.

The National Ocean Service has identified eutrophication—the overabundance of nutrients and minerals in a body of water—as a major threat to the nation's waterways, but nutrient pollution is a global problem that does not respect national borders. Its impacts are everywhere from the Great Barrier Reef off the coast of Australia to the huge dead zone outside the Mississippi River delta. It leads to the overgrowth of cyanobacteria and toxic algae blooms, fish kills, and dead zones, as animals are starved of oxygen.[28]

Of course, poisoned waters harm many forms of marine life, not just fish. Phosphate is a major source of eutrophication.[29]

While phosphate fertilizers are an obvious source of phosphate, there is another ominous source that is being mostly overlooked: glyphosate. Glyphosate itself is water soluble, so it can freely enter the waterways where it can be metabolized by any microbes that are able to break it down. While many microbes are unable to metabolize glyphosate, cyanobacteria love glyphosate, exploiting it as a source of phosphorus to produce phosphate. Cyanobacteria flourish in glyphosate-contaminated waters.[30] This is a huge and growing problem in the shallow ocean waters surrounding agricultural regions where glyphosate is heavily used, including in and around the Florida coast. Scientists believe that heavy rains, flooding, and flood plumes, not to mention rivers, streams, and other waterways, can deliver dissolved glyphosate far from shore into the ocean where it then persists for an extended period with little degradation. Glyphosate persists in seawater in the absence of sunlight, lasting for up to 315 days in the dark at 31°C. It will persist for 47 days under the more realistic conditions of low light at 25°C.[31]

The problem of glyphosate in waterways isn't limited to eutrophication, however. There's also the impact of glyphosate itself on aquatic life. Daphnia is a water flea that is a common inhabitant of freshwater habitats worldwide, where it's central to the aquatic food web. But scientists from laboratories in the UK and the US recently found that both glyphosate and Roundup increase the number of aborted eggs and juveniles born dead. Both glyphosate and Roundup also delays maturation and decreases the size of the offspring at birth, induces DNA damage, and disrupts the gut microbiome of the water fleas. This research, published in 2020, also revealed that a species of microbes known to be able to fully break down glyphosate is over-represented in the fleas' microbiome. Due to daphnia's essential role in the ecosystem, the study authors warn that "the weedkiller can potentially impose a fitness burden on freshwater aquatic foodwebs." I would argue it already does.[32]

Scientists have also found that crabs exposed to glyphosate suffer from reduced sperm counts and increased numbers of abnormal sperm.[33] In a thorough and carefully designed study on freshwater crayfish exposed to subacute concentrations of glyphosate, scientists looked for changes in the activities of various enzymes in the hemolymph (their equivalent to blood) compared to controls. Crayfish are very sensitive to chemical exposure. The glyphosate exposure caused cell damage in the liver, pancreas, and hemocytes (blood cells). Several enzymes, used mainly in the liver and pancreas, were reduced in activity, indicating damage to the liver and pancreas. The shells of the glyphosate-exposed crayfish were also unusually soft.[34] The damage these scientists saw to the blood cells and the increased oxidative stress have been linked to immune dysfunction.[35]

Perhaps the most disturbing aspect of this study, however, was the *behavior* of the crayfish. Prior to exposure, the crayfish congregated in social groups to share food. After exposure they "quarreled with each other to take shelter."[36] This behavior was not observed in the control group. Might this be due to serotonin deficiency induced by glyphosate's disruption of the shikimate pathway in the gut microbiome? Serotonin is produced from tryptophan, one of the three amino acids synthesized by plants and gut microbes via the shikimate pathway. Glyphosate's blockage of the shikimate pathway is believed to be the primary mechanism by which it kills plants. Most of the serotonin produced in the body is produced in the gut, and deficiencies in tryptophan in glyphosate-exposed food sources combined with impaired tryptophan synthesis by gut microbes could conspire to produce a systemic serotonin deficiency, which has been linked to aggressive, violent behavior.[37]

I'll give you another stunningly sad example. Stranded dolphins are showing up with amyloid beta plaque in their brains, a characteristic sign of Alzheimer's disease. In 2019, a team of marine biologists, neurologists, chemists, and even a medical examiner examined 14 dolphins that had been stranded in Florida and Massachusetts. The brains of 13 out of the 14 of the dolphins were found to contain abnormally high levels of β-N-methylamino-l-alanine.[38] This amino acid, found in 93 percent of these dolphins' brains, is produced by cyanobacteria and is known to be neurotoxic.

There is no life on Earth that can survive without water. When we poison waterways we poison ourselves.

The Last Butterfly

As a child, I loved to be outside in the evening to watch in wonder as the bioluminescent fireflies twinkled in the darkness. There are more than 200 million insects for every human on the planet. Insects massively outrank all other animals in terms of diversity, sheer numbers, and total biomass. It's estimated that 80 percent of wild plants depend on insects for pollination, and 60 percent of birds rely on insects for food. Preserving the abundance and diversity of insects is arguably as important as keeping the waterways free of pollution.

But we seem to be fighting a losing battle. The number of insects around the world has been declining rapidly, and a full 40 percent of insect species are currently threatened with extinction.[39] When German scientists monitored insect populations across 96 locations in various protected areas, they found a 76 percent decline in flying insect biomass between 1989 and 2016.[40] Moths, butterflies, wasps, bees, ants, and dung beetles seem to be the most affected. Pollinating insects, especially, have experienced steep declines.

The monarch butterfly, a once commonplace iconic symbol in North America, is on the verge of extinction. The monarch makes an epic flight each fall from the northern plains of the United States south to Mexico City, and then makes a return trek in the spring. There has been a decline of more than 80 percent in the monarch population in the East, and an ominous decline of 99 percent in coastal California from an estimated 4.5 million in the 1980s to only 28,429 today.[41]

Several factors are to blame: habitat loss from converting land to agriculture, predators that kill monarchs faster than their populations can regenerate, climate change, and pesticides, including glyphosate. Empirical evidence shows an inverse relationship between glyphosate application rates by county and local monarch populations.[42] The main food that the monarch caterpillar eats is milkweed, a common weed growing among GMO Roundup Ready corn and soy crops. While the loss of milkweed has been proposed as a causal factor in the decline in monarch butterflies, what has not yet been recognized is that butterflies are also being poisoned by glyphosate in the milkweed that they do find.

Butterflies aren't the only insects being decimated. When water fleas (genus *Daphnia*) are fed Roundup Ready soy containing glyphosate residues from commercial farm harvests, they experience stunted growth, have fewer offspring or are unable to reproduce altogether, and die in record numbers.[43] Though negative health effects are more pronounced at the higher levels of exposure, the fleas' health outcomes are also poor at levels well below maximum residue limits set by the EPA.

We've also seen an alarming decline in the honeybee population over the past two decades. Environmentalists, naturalists, farmers, and researchers are sounding the alarm about colony collapse disorder, a catastrophic phenomenon that occurs when the majority of the worker bees suddenly disappear, abandoning the beehive's larvae and immature bees, causing a collapse of the colony. Humans may not be able to survive without bees. These buccaneers of buzz, as the nineteenth-century poet Emily Dickinson called them, are master pollinators. They fertilize food crops and play a crucial role in wild plant diversity. Humans aren't the only animals who get food from bees. Bears, possums, raccoons, skunks, and insects will eat honey, honeycomb, larvae, and the bees themselves. Bees are a food source for dozens of different birds, as well as dragonflies, spiders, and praying mantises. Pathogens, parasites, genetics, climate change, and a loss of foraging habitat have all played a role in the decline of the honeybee.[44] Neonicotinoid insecticides, especially Imidacloprid, have been identified as a main culprit.[45]

Neonicotinoids are a class of insecticides used in agriculture for seed and soil treatment, as well as in home gardens and on golf courses to control unwanted insects like aphids, whiteflies, and thrips. They are also used as a flea and tick treatment for pets. These insecticides disrupt the central nervous system, killing insects when they come into contact with them or ingest them. When bee larvae are exposed to neonicotinoids, they show changes in gene expression. For example, scientists have found a sharp increase in the production of Cytochrome P450 (CYP) enzymes in exposed bees. CYP enzymes are important for detoxification of many fat soluble toxic chemicals and toxins.[46] Glyphosate has been shown to suppress liver CYP enzymes in both rats and chick embryos.[47] Neonicotinoid detoxification depends on CYP enzymes, which glyphosate suppresses. This suggests that simultaneous exposure to neonicotinoids and glyphosate may be more devastating than either would be alone, and that scientists should be looking carefully at synergistic effects of the two pesticides together.

One of the most important functions of honeybee colonies is to fertilize flowering plants. Among them are almond trees. Almond trees have five-petaled white or light pink flowers that bloom in early spring. Presumably since the trees bloomed so early, for early Christians, almonds symbolized resurrection. In China, almond cookies that look like coins are a favorite treat during New Year celebrations, symbolizing good fortune. But billions of bees are dying after being in almond groves, which are commonly treated with glyphosate. A recent survey of beekeepers showed more than a third of commercial bee colonies, some 50 billion bees, died over the winter of 2018–2019.[48]

What do scientists find when they examine glyphosate's effects on bees? Nothing good. When adult worker honeybees are exposed to sublethal doses of glyphosate, it decreases their short-term memory retention and disrupts the associative learning necessary for effective foraging.[49] Bees are remarkable insects. They recognize each other's faces and they can even be taught to recognize humans. Healthy bees, with practice, find the shortest way back to their hives. But bees exposed to seemingly "safe" levels of glyphosate take longer to return to their hives, and they are not able to improve their time spent finding their way back home in a second release.[50] Even when it does not kill them outright, glyphosate damages their ability to find food. In 2019, a team of scientists from Argentina examined changes in foraging behavior of honeybees exposed to glyphosate in a laboratory setting.[51] They discovered that glyphosate negatively affects their learning, cognitive, and sensory abilities.

Glyphosate also disrupts the honeybee microbiome.[52] *Snodgrassella alvi* is an important species in the bee microbiome. All strains of this species encode a version of the enzyme EPSP synthase that is sensitive to glyphosate. When worker bees were exposed to small amounts of glyphosate added to sucrose syrup for 5 days, they had lower numbers of the beneficial *S. alvi*, *Bifidobacterium*, and two *Lactobacillus* strains. At the same time, the glyphosate exposure made the bees more susceptible to the pathogenic bacteria *Serratia marcescens*. Bees with impaired microbiomes are more likely to abandon their hives and die from opportunistic infections.

Worms Struggling Blindly

Ramona Quimby, the exuberant troublemaker in Beverly Cleary's beloved novels, plucks an earthworm out of the soil to give to a boy as an engagement ring. Children, farmers, and fisherman all love earthworms, as do blackbirds, godwits, black-bellied plovers, robins, woodcocks, and other worm-eating birds. Earthworms play an essential role in maintaining healthy soil. They are ecosystem engineers, shredding plant litter and mineralizing it in their guts, breaking down complex nutrients into simpler ones that are more easily absorbed by plants. Their manure castings are excellent fertilizer and their below-ground burrowing improves root penetration and water filtration. A single square meter of land can carry up to 1,000 earthworms.

Scientists in Ireland, however, report a decline in earthworm populations in tilled soil.[53] A recent study in Great Britain also found a reduction in earthworms on many farms.[54] The decline in earthworms may be the main cause of the alarming declines in song thrushes, one of Britain's most cherished birds.[55]

In 2015, a team of scientists in Austria conducted a revealing experiment.[56] The team set up pots seeded with grasses and herbs in a greenhouse and introduced two distinctly different species of earthworm: one that burrows vertically and another that stays underground. The researchers applied just over half the amount of Roundup recommended for agricultural use and then carefully collected data on the worms' activity, compared to the controls. The scientists anticipated that the burrowing worms would produce more surface castings following herbicide treatment than in the control, predicting that the availability of decaying plant material would attract the burrowing earthworms to the surface.

The opposite turned out to be true. After just one week, the casting rate of all earthworms exposed to glyphosate, both the burrowing and the non-burrowing, dramatically decreased. After three weeks, their casting virtually stopped; the exposed earthworms had essentially stopped eating and excreting. We don't know why. Perhaps the earthworms rejected the decaying plant matter because they could detect glyphosate and refused to eat it. The more likely hypothesis is even more disturbing: that they developed impaired mobility due to glyphosate disrupting enzymes in their muscles.[57] A third possibility is that the glyphosate damaged their nervous systems. This is certainly plausible given that a 2012 study on roundworms found glyphosate caused damage to the worms' neurons that release dopamine, resulting in a condition similar to Parkinson's disease.[58] Reproduction of both species of earthworm was greatly impaired, reduced by 56 percent in the soil-dwelling worms three months after the herbicide was applied.

Silent Marshes

I have seven brothers and sisters. Frogs were abundant in the marshes near our childhood home in Connecticut. One of my brothers spent many hours of pure joy trying to catch them, bringing them home, and inviting us to revel in their cold squishiness with him. Like earthworms, amphibians are cold-blooded animals with moist skin. Unlike earthworms, amphibians go through two phases of life. Amphibian larvae, called pollywogs or tadpoles, live in the water and use gills to breathe. Then, in an incredible morphological feat, they metamorphose, losing their tiny tails and gills and growing limbs and lungs.

Frogs, toads, salamanders, and newts are all amphibians and have been studied more extensively than earthworms. Many recent findings aren't good. Amphibians have suffered from major declines due to severe diseases over the past few decades. A single chytrid fungus has led to the decline of at least 501 amphibian species and some 90 extinctions.[59] Forty-one percent of known amphibians now risk extinction.[60] We know that amphibian populations are threatened by invasive predators, parasites, climate change, emergent diseases, and degraded habitats. What is less well known is that amphibians are especially susceptible to pesticides like glyphosate. Amphibians don't drink water; they absorb it through their skin. This highly absorbent skin makes frogs, toads, newts, and salamanders vulnerable to environmental contaminants.

In 2005, scientists discovered that when outdoor ponds and surrounding areas were sprayed with Roundup, 96 to 100 percent of the tadpoles died in three weeks. Incredibly, 79 percent of juvenile frogs and toads on land died *in just 1 day*.[61] In 2019, French scientists exposed the eggs of African clawed frogs to glyphosate and examined the developing embryos for developmental defects. A remarkable number of embryo defects indicated severely crippled processes involved in DNA replication and cell division, the first step when a fertilized egg divides to produce a mature embryo. These defects included disorganized spindles, disorganized chromosomes, ectopic spindles, and missing spindles or chromosomes. Researchers discovered "double spindles" and "double asters" that had never before been recorded in toxicology studies. Glyphosate also delayed the maturation process.[62]

A 2019 study on tadpoles showed synergistic toxicity between glyphosate and arsenic, both of which are present in the ground water of the Argentine

Chaco-Pampa Plain.[63] The scientists found a significant increase in thyroid hormone synthesis, which can disrupt development, when both chemicals were present. They noted that the combination of glyphosate and arsenic is a potent endocrine disruptor, causing more DNA damage in the tadpoles than by the effects of either chemical alone.

Harm to Our Pets

Domestic animals are also being exposed to high levels of glyphosate.[64] Researchers based in New York State found widespread occurrence of glyphosate and its derivative, AMPA (aminomethylphosphonic acid), in the urine of dogs and cats. N-methyl glyphosate (a glyphosate molecule with a methyl group (CH_3) attached to its nitrogen atom) was also detected in the urine. Both glyphosate and N-methyl glyphosate are carcinogenic; N-methyl glyphosate may actually be more carcinogenic than glyphosate itself.[65] "Grain-free" dog foods, often based on legumes such as lentils, chick peas, and fava beans as a source of protein, are becoming increasingly popular. The highest levels of glyphosate are consistently found in legumes. Pets are being overexposed to glyphosate in at least three ways: by eating it (herbicide residues in their pet food), drinking it (herbicide residues in drinking water), and getting it on their skin or in their lungs via lawns, public parks, and home gardens.

In July 2018, the FDA released a statement warning dog owners about cardiomyopathy. Canine dilated cardiomyopathy is a relatively new disease among dogs and is believed to be caused by their increasingly toxic diets.[66] A similar heart malady has long been recognized in cats. Interestingly, increasing the content of taurine in cat food has greatly reduced the risk, and taurine supplementation also seems to improve symptoms in dogs with cardiomyopathy. Taurine is stored in high concentrations in the heart and may be crucial for renewing sulfate supplies to the heart following a heart attack. How is this related to glyphosate? Glyphosate severely suppresses taurine uptake.[67] We'll talk more about taurine in chapter 6.

Other harmful additives in pet food may compound the problem of glyphosate toxicity. Purina brand cat food products contain ethoxyquin, an artificial preservative developed by Monsanto, that has been shown to damage the liver and kidneys and that is actually banned from use in all food for human consumption (except spices). While pet food companies continue to add it even to so-called "natural" products in America,[68] in 2017 the European Union suspended its use in all animal feed.[69]

In one of my favorite songs, Peter, Paul, and Mary ask where all the flowers have gone. The flowers have all been picked by the young girls who have laid them at the graves of their young men. The men are also gone, soldiers who have died in battle. Pete Seeger wrote that song, which he taped to a microphone and sang for the first time at Oberlin College in 1955, to protest the senseless losses that come from war.

Today we're fighting a different kind of war. Without worms to churn and nourish the soil and without bees and butterflies to pollinate them, flowers cannot grow. From insects to birds to our four-legged mammalian friends, the living world is suffering from environmental pollutants. From increased virulence of fungi to the alarming collapse of some species, a disrupted ecosystem, and poisoned waterways, it's no exaggeration to say that glyphosate and other human-made toxicants, if we continue to use them in this unchecked and irresponsible way, are destroying the natural world.

— CHAPTER 3 —

Glyphosate and the Microbiome

I submit to you that in any disease that has an immune component—whether it's Alzheimer's, Parkinson's, autism, atherosclerosis, obesity, diabetes and any disease that you are seeing in your clinics with an immune component—the microbiome is having some effect.

—James T. Rosenbaum, MD,
Oregon Health Sciences University

Glyphosate is considered so safe that no US governmental agency tests for it in food. Many homeowners use it without a second thought to control weeds in their yards. It's in the rain, the water supply, and the cotton in clothing, dish cloths, tampons, and baby diapers. It aerosolizes, becoming dispersed into the air, which means we breathe it into our lungs. And it's present in much of what we eat. Even if you eat an organic diet, as I mentioned earlier, it's likely you're still being exposed to glyphosate.

The companies that manufacture and sell glyphosate are so powerful that they have convinced government authorities and many research scientists that glyphosate is so safe that they don't even bother to investigate whether it could be partly to blame when bees and butterflies start dying in alarming numbers. At the same time, we all know that pollution is harming the plants and animals we humans rely on.

When scientists look at time trends in the United States for multiple diseases that are becoming more prevalent in the past few decades, and compare them to the time trends for the use of glyphosate over the same time period, they find stunning correlations between the rise in glyphosate use and the rise in Alzheimer's disease, autism, diabetes, inflammatory bowel disease, kidney disease, liver disease, obesity, pancreatic cancer, and thyroid cancer.[1]

Some dismiss the results of this research with the claim that "correlation does not equal causation." Fair enough. But consider this: A p-value is a mathematical measure of the probability that a correlation between two curves could have occurred by chance. A p-value of 0.05 is considered significant. As Nancy Swanson and her team of scientists explain, "When correlation coefficients of *over 0.95* (with p-value significance levels less than 0.00001) are calculated for a list of diseases that can be directly linked to glyphosate, via its known biological effects, it would be imprudent not to consider *causation* as a plausible explanation" [my emphasis].[2] If exposure to glyphosate is indeed causing the decline in human health in industrialized countries, as I believe it is, we need to ask ourselves how.

Disrupting a Microbial Pathway

Glyphosate's effect on plants is to disrupt the shikimate pathway, a metabolic pathway that plants use to produce the aromatic amino acids tryptophan, tyrosine, and phenylalanine, which are the precursors to proteins, vitamins, and other kinds of bioactive substances such as pigments, hormones, and neurotransmitters. When glyphosate disrupts the shikimate pathway, it kills the plants. This makes glyphosate an extremely effective herbicide. How other organisms are affected by disruption of the shikimate pathway is a subject of contentious debate.

Monsanto researchers and other industry-funded scientists posit that, because the shikimate pathway does not exist in human cells, glyphosate doesn't pose a risk for us. As you now know, however, many of the microbes in our bodies *do* possess the shikimate pathway. Not all bacteria are equally sensitive to glyphosate. A study by researchers in Finland used a bioinformatics approach to predict which microbial species in the gut would be sensitive to glyphosate toxicity. They found that 54 percent of the species found in the gut carry a glyphosate-susceptible version of EPSP synthase.[3] So while humans can derive these three aromatic amino acids from diet, the bigger issue is this: glyphosate kills anything that *possesses* the shikimate pathway, and that includes much of our microbiome—microbes we rely on for providing us with nutrients, aiding digestion, maintaining a healthy gut barrier, and promoting the development of a healthy immune system.

For example, a genomic assessment of the human microbiota demonstrates that microorganisms collaborate to produce the B vitamins: thiamine (B_1), riboflavin (B_2), niacin (B_3), pantothenate (B_5), pyridoxine (B_6), biotin (B_7), folate (B_9), and cobalamin (B_{12}). It further reveals that these B vitamins made inside the body significantly augment B vitamins supplied from food. Our human cells are unable to produce these vitamins. But, collectively, our microbial hitchhikers contain a large number of enzymes that specialize in various steps in the synthesis of these essential nutrients, and many of these microbes are dependent on a working shikimate pathway.[4]

In plants, fungi, and bacteria, glyphosate's attack on the shikimate pathway is mainly localized to an enzyme called EPSP synthase. A synthase is an enzyme that links together two molecules to make a third molecule, its product. In the case of EPSP synthase, the two molecules are phosphoenolpyruvate (PEP) and shikimate 3 phosphate (S3P). EPSP synthase pries a phosphate loose from PEP, and this creates energy that allows it to "sew" the remaining piece together with S3P to create 5-enolpyruvylshikimate-3-phosphate (EPSP). The phosphate is important, because it's the breaking of the high-energy phosphate bond that energizes the reaction. And, as we will see later, glyphosate interferes with the binding of EPSP synthase to PEP at the phosphate site.

EPSP is an intermediary in the shikimate pathway, not the end result. EPSP is further processed by other enzymes to achieve the primary goal of the pathway, which is production of the three aromatic amino acids: tryptophan, tyrosine, and phenylalanine. Human cells do not have any of the enzymes of the shikimate pathway, including EPSP synthase. For this reason, our body depends on our food sources and on our gut microbes to produce these amino acids for us. This is why they're called *essential* amino acids. These amino acids are not only essential for building proteins. They are also the building blocks of many other molecules that play a critical role in our biology, including many of the B vitamins; the neurotransmitters serotonin, melatonin, dopamine, and epinephrine; thyroid hormone; and the skin-tanning agent melanin.

Glyphosate has been shown experimentally to block PEP from binding to EPSP synthase, specifically by getting in the way of the phosphate attached to PEP. For the moment, I want you only to be aware of this phenomenon. We will return to it in chapters 4 and 5, because it has enormous significance in predicting which other enzymes would be affected by glyphosate through an analogous mechanism.

Teeming with Nonhuman Life

We think of ourselves as humans, which we are. But each of us is teeming with nonhuman life, microscopic organisms that live on us and inside us. Scientists are beginning to understand that this microbiome, as our nonhuman cohabitants are collectively called, plays a seminal role in human health. The gut microbiome is a teeming collection of *trillions* of bacteria, viruses, and fungi that have made the human gut their home. Their relationship with us humans is largely symbiotic: They perform many functions for us, producing all sorts of biologically useful molecules that the host cells are unable to synthesize on their own. There are estimated to be 10^{14} microbes residing in the human gut alone. While it has been widely reported that microbial cells outnumber human cells by a factor of 10, careful analysis suggests that they may match our own cells "only" one-to-one.[5] Nonetheless, it is undisputed that their collective genome carries more DNA code than ours by a factor of at least 100.[6]

For most of the last century, the gut microbiome was pretty much ignored in research on the human body. I suspect this is partly because it was functioning well. We did not notice all the things the microbes were doing for us because we did not need to. This is no longer the case. From colicky babies who are unable to digest their food to my friend's 27-year-old sister who cannot eat anything without getting a stomach ache, our guts are more inflamed and we are in more pain than ever before.

When we disrupt the gut we also risk disrupting the brain. Scientists now understand that the gut and the brain are in close communication. The signaling that takes place between the gastrointestinal tract and the central nervous system is called the gut-brain axis. Communication happens via the lymph system, via the blood circulation, and via the vagus nerve.[7] A lot of this communication involves signals released by the microbes in the gut.[8] This is why several of our modern diseases are now believed to have their origins in the gut, including Alzheimer's disease, amyotrophic lateral sclerosis (ALS), autism, depression, Parkinson's disease, and rheumatoid arthritis, among others.

Gut microbes are essential for promoting the generation of neurons in the hippocampus, which plays a central role in brain development, or neurogenesis. Antibiotics can have a profoundly negative effect on this process. In a study on mice, antibiotic treatment led to a reduced number of monocytes, a type of immune cell, in the brain. Intragastric treatment of mice with antibiotics caused clear deficits in brain function, most especially in the ability to recognize novel objects when they are reintroduced, and it was associated with particular changes in neurochemical brain activity. The researchers proposed that the cognitive impairment was specifically caused by gut dysbiosis.[9] These results are disturbing: They suggest that antibiotics can damage the brain.[10]

Each person's microbiome is unique, with many different species coinhabiting the gut in a symbiotic relationship with one another as well as with us. Which species take hold depends in part on the microbiome of the mother at the time a child is born. We know that C-section birth can disrupt the microbiome, leading to microbes in the child's gut that more closely resemble the species that normally inhabit the skin. This can set up the child for a rough start in terms of achieving a proper balance in the gut.[11] We also know that breast milk nourishes the *Lactobacillus* species that normally thrive in the infant gut. In particular, *Lactobacillus casei* feasts on the lactose in milk.[12] These are more than just "friendly" bacteria. These are symbionts that play an essential role in digesting our food, protecting us from disease, and keeping us healthy. When a baby starts eating solid foods, the microbiome changes dramatically to accommodate the rich variety of nutrients in the newly introduced foods.

The majority of children with compromised brain function, including autism and autism-like disorders, suffer from gut problems.[13] An inflammatory gut and a leaky gut barrier allow pathogens and toxic microbial metabolites to enter general circulation, which can cause a systemic inflammatory response, including inflammation in the brain.[14] Chronic low-grade encephalopathy, or inflammation of the brain, is also associated with mood disorders and cognitive problems.

For example, scientists have found that children with autism have more *Clostridia* in their gut, as well as different strains of *Clostridia*, than children without autism.[15] There's also a strong association between certain toxic metabolites produced by *Clostridia* and encephalopathy. *Clostridia* are less sensitive to glyphosate than other gut microbes like *Lactobacillus* and *Bifidobacteria*, leading some scientists, including me, to postulate that glyphosate is causing a microbial imbalance leading to the production of toxic metabolites that contribute to brain damage.

We can actually induce brain damage in animals through a process called maternal immune activation.[16] In 2013, scientists found that maternal immune activation during pregnancy was associated with behaviors characteristic of autism in the offspring. The distribution of gut microbes in the brain-damaged mice was also striking: These mice were deficient in *Bacteroides fragilis* but replete with *Clostridia* species that were producing a metabolite called 4-ethylphenylsulfate (4EPS). The 4EPS metabolite was found to be 46-fold higher in the autistic mice than in controls. This metabolite is similar to *p*-cresol, which is known to be elevated in humans with autism.[17]

Treatment of the mice with a probiotic enriched in *B. fragilis* led to improvements in autistic symptoms, reduced expressions of anxiety, and produced drops in 4EPS blood levels. At the same time—and this is an astonishing finding—the scientists were able to *induce* anxiety in control mice, simply by exposing them to 4EPS. Brain health and immune system health are intricately intertwined. An unhealthy gut microbiome compromises both. Remarkably, the species *B. fragilis*, the deficiency of which is tied to autism, has also been shown to protect the host from viral infections.[18]

We know, without doubt, that a healthy gut microbiome—in both humans and mice—during the first few weeks of life is crucial for brain health. Mice with unhealthy microbiomes have a more acute stress response than normal mice. They show increased elevation of plasma adrenocorticotropic hormone (ACTH) and corticosterone, both indicators of stress. But when the guts of these mice are reconstituted with *Bifidobacterium infantis* early on, their exaggerated stress response disappears. If improving their unhealthy microbial balance is delayed to a later stage in development, brain problems remain.[19]

A Barrier Easily Breached

A crucial aspect of both gut health and blood health is the balance of acidic and basic influences on the pH. The term *pH* is used in chemistry to characterize the prevalence of protons (H^+) and hydroxyl ions (OH^-). A pH of 0 is strongly acidic (many protons). A pH of 14 is strongly basic (many hydroxyls). Measurements of pH are made in a range from 0 to 14, so a pH of 7 is exactly neutral, with equal numbers of protons and hydroxyl ions. The human body has complex mechanisms to maintain the proper pH of the blood and of the gut lumen. Blood is normally slightly basic, with a pH of 7.3–7.4. Organic molecules influence the pH of the water in the gut. For example, acetate is acidic; ammonia is basic. If there's too much ammonia and too little acetate in your intestines, the pH of the gut will become higher. Changes in pH have a complex effect on the microbial and metabolic activities in the gut.

Even as far back as the 1970s, researchers knew that acid-loving species such as *Lactobacillus* and *Bifidobacteria* are important to a healthy gut. Long before talking about the microbiome was in vogue, in 1973, scientists were speculating that increasing the growth of acidophilic bacteria such as *Lactobacillus* and *Bifidobacteria* "could depress the growth of putrefactive, ammonia-producing organisms such as *E. coli*."[20] Our health improves when we have more *Lactobacillus* and *Bifidobacteria* in our bodies, two classes of bacteria most sensitive to glyphosate.[21]

There is a strong correlation between the pH of feces and the amount of glyphosate detected in the colon.[22] There is also a strong *inverse* correlation between the amount of acetate in the cecum and the amount of glyphosate in the cecum. This suggests that glyphosate interferes with microbial production of acetate and that this may be a contributor to high pH in the gut. Glyphosate causes the pH of the gut to go up and levels of acetate to go down, most likely because it is disrupting the synthesis of acetic acid by gut microbes. This is important because acetic acid is a precursor to acetyl coenzyme A, a molecule that feeds into the citric acid cycle to generate energy for the cell.

Our bodies make specialized enzymes that help digest our food. Glyphosate can infiltrate these digestive enzymes. So another cause of high pH following chronic glyphosate exposure is glyphosate infiltration into digestive enzymes. Anthony Samsel and I reported remarkably high levels of contamination of glyphosate in the digestive enzymes trypsin, pepsin, and lipase, sourced from pigs.[23] Glyphosate may be disrupting the ability of trypsin and pepsin to digest proteins, as well as the ability of lipase to digest fats. With impairments in trypsin and pepsin, undigested proteins can make their way into the colon, where they are broken down by the gut microbiome, releasing ammonia.[24] As I explained previously, ammonia is extremely basic, which means that it drives the colonic pH way up. Acetate (acetic acid), as its name implies, is acidic, so it lowers the pH. But acetic acid is reduced by the influence of glyphosate.

The surface of the gastrointestinal tract is made up of a single layer of tightly interconnected epithelial cells. These cells are covered with mucus, a complex biological material that forms a barrier to protect cells from damage. Mucus, which consists of highly sulfated glycoproteins called mucins, allows passage of gases, nutrients, and many proteins.[25] Under healthy conditions, mucus forms an excellent barrier that keeps the surface layer of cells safe from attack by any potentially damaging products in the intestines. The mucins in the colon are especially dense in sulfate, which is thought to protect them from degradation.[26]

A major component of human breast milk is a complex glycoprotein rich in sulfated mucins, including heparan sulfate, chondroitin sulfate, and dermatan sulfate.[27] These glycoproteins, collectively called oligosaccharides, contain large amounts of complex sugars that human cells can't metabolize. *Bifidobacteria*, and especially *Bifidobacterium infantis*, specialize in metabolizing the sulfated mucins found in human milk. Once detached from the parent protein, sulfated mucins become available to bind to the infant gut wall and help maintain a healthy gut barrier. *B. infantis* metabolizes the complex sugars in the oligosaccharides into lactate and acetate, and, in doing so, lowers the pH of the gut.

In 2018, a team of scientists examined trends over time of the pH of the infant gut, going back to the 1920s.[28] The researchers' working hypothesis was that something in the modern environment was causing *Bifidobacteria* to become less able to dominate the infant gut microbiome, particularly in wealthier countries. In 2021 scientists revealed that *B. infantis* strains with a complete capacity to metabolize milk oligosaccharides are now exceptionally rare in the guts of infants in the United States.[29] In 1913, the infant gut was an "almost pure culture" of *Bifidobacterium*.[30] Today, studies have found that the microbial mix in the infant gut is much more diverse and that infants excrete *undigested* human milk oligosaccharides in their feces in large amounts.[31] Prior to 1980, studies showed a fecal pH that was less than 5.5, whereas after 1980 the pH had a value greater than 5.5, with the highest values (up to 6.5) appearing after 2000. Scientists hypothesize that this pH change is due to a dramatic reduction in *Bifidobacteria* species in the human gut in recent decades.

Glyphosate was introduced into the food chain in 1975, and genetically engineered Roundup Ready crops were introduced in the mid-1990s. *Bifidobacteria* are among the most sensitive to glyphosate of all microbes that have been studied.[32] Scientists have found that maternal glyphosate exposure is associated with reduced *Bifidobacteria* in the milk as well as with increased risk of premature and difficult birth.[33] A study on pregnant women in Indiana found an association between premature birth and glyphosate levels in the urine.[34] Loss of *B. infantis* in breastfed infants leads to chronic enteric inflammation during a critically important immunological training window.[35] The increase in C-section deliveries, the increased use of antibiotics, as well as the increased practice of formula-feeding also contribute to the loss of *Bifidobacteria*.[36]

Humans are caught in a vicious cycle. Most infant formula is contaminated with glyphosate. Soy formula from Brazil was found to have levels of glyphosate up to 1,000 parts per billion.[37] Exposure to glyphosate from soy formula reduces the presence of *Bifidobacteria* and interferes with the process by which these bacteria maintain a healthy gut pH and a healthy recycling of mucins lining the gut barrier, as well as a steady supply of acetate to fuel the mitochondrial production of energy in the form of adenosine triphosphate (ATP). Many proteins rely on the last phosphate in ATP to generate usable energy to support enzymatic reactions, muscle contraction, ion transport, and other activities. Feeding soy formula to an infant is a triple whammy: The soy itself is an endocrine disruptor, the formula does not contain breast milk's rich glycoproteins that maintain healthy mucins and support the growth of *B. infantis*, and the glyphosate in the soy disrupts the microbiome, further weakening its protective barrier.

Bacteria Fight Back: Microbial Strategies to Resist Glyphosate

One factor that influences the balance and health of the microbiome is the degree to which different species of bacteria can protect themselves from glyphosate. Bacteria have developed several clever strategies for acquiring resistance to glyphosate. Some are naturally resistant. Their version of EPSP synthase is not adversely affected by glyphosate. For example, *Staphylococcus aureus* (MRSA), which has become a widespread problem in hospitals, possesses a form of EPSP synthase that is insensitive to glyphosate.[38] Other species with a more sensitive version of EPSP synthase have evolved to vastly overproduce EPSP synthase in order to compensate for its low activity. Other bacteria are even able to break down glyphosate through specialized enzymes and use it as a source of nutrients.

Another strategy bacteria use is to modify genes that carry glyphosate into the cell to prevent the glyphosate from gaining entry. It appears that glyphosate can gain entry into cells by hitching a ride on a protein that carries the amino acid glutamate into the cell. Curiously, the soil bacterium *Bacillus subtilis* has evolved with a mutated and dysfunctional form of the gene that codes for the protein that imports glutamate. This blocks glyphosate uptake by the cell.[39] This finding suggests that glyphosate can enter cells along glutamate transport channels. Glutamate and glyphosate are about the same size, and both molecules are negatively charged amino acids.

Other microbes have developed resistance by perfecting genes that can metabolize glyphosate. This is an "ideal" solution from the microbe's perspective, because it destroys the glyphosate in the process. For example, a strain of the infection-causing bacteria *Pseudomonas* is among the very few known species that can fully metabolize glyphosate. This could be a factor in *Pseudomonas aeruginosa*'s emergence as a major problem in hospitals.[40] On the other hand, a microbe that can metabolize glyphosate may well benefit its host by clearing glyphosate from the body. We simply don't know. We don't understand enough about the implications of tinkering with biological systems in this way.

Powerful Pathogens

Antibiotics such as penicillin have been a boon over the past century in curing infections that were once untreatable. In 1939, Dr. Ernst Chain, a

German-born British chemist, injected eight mice with a virulent strain of *Streptococcus*. He then injected four of the mice with penicillin and left four without treatment. In the morning the treated mice were alive, and the others were all dead. It's no wonder that Dr. Chain and generations of people to follow hailed penicillin as a miracle.[41]

Unfortunately, this miracle has also led to an unexpected problem. In the past few decades, powerful pathogenic bacteria that are resistant to antibiotics have become a looming threat. These include multidrug-resistant *Pseudomonas aeruginosa*, drug-resistant *Salmonella*, methicillin-resistant *Staphylococcus aureus* (MRSA), drug-resistant *Streptococcus pneumoniae*, and vancomycin-resistant *Enterococcus*, among many others.[42] Antibiotic resistance is a global concern, one that has already given rise to lethal untreatable infections and is predicted to give rise to many more.

Scientists have learned that chronic exposure to one antibiotic can allow pathogens to develop broad resistance to others. Remember: Glyphosate is an *antibiotic*. It was patented by Monsanto in 2010 for use as an *antibiotic* to control microbial infections. When we are chronically exposed to glyphosate in our food and water, it is like taking low doses of antibiotics over an extended period of time. While glyphosate improves the effectiveness of some antibiotics, it has the opposite effect for others. In particular, glyphosate *reduces* the responsiveness of both *Escherichia coli* and *Salmonella typhimurium* to ciprofloxacin (Cipro) and kanamycin, two commonly used antibiotics.[43] The effects of concurrent exposure to low-dose glyphosate on various antibiotics can be catastrophic.

In 2019, an international team of scientists discovered that antibiotics interfere with a signaling mechanism in the lungs that launches the first immune response to flu infection.[44] Mice given antibiotics have a worse outcome when infected with the influenza virus. However, and this is significant, a fecal transplant from a mouse that has not been exposed to antibiotics restores gut health and increases lung resilience against the flu in a mouse that has been infected with the influenza virus. This finding aligns with the study mentioned earlier that showed that *Bacteroides fragilis* protects the host from viral infections.

What does this mean? Chronic exposure to glyphosate, an antibiotic, may make humans more susceptible to the flu and other respiratory infections, including COVID-19.

Bowels on Fire

Gut microbes in the colon break down complex carbohydrates that escape digestion in the middle intestine. These carbohydrates are known as prebiotics, and bacteria convert them into short chain fatty acids, primarily acetate, propionate, and butyrate. The balance among these three fatty acids has important implications for gut health, and is strongly influenced by the pH of the gut.[45] In particular, butyrate is important for maintaining a healthy gut barrier, since it is a major nutrient for the epithelial cells (called colonocytes) that line the surface of the colon. It's worth noting here that researchers have been able to induce autism-like behaviors in mice by exposing them to excess propionate.[46] (I return to the subject of propionate and its relationship to autism in chapter 9.)

A study of gut bacteria under controlled pH conditions revealed a remarkable pattern consistent with glyphosate's effects on pH and microbial populations.[47] A low pH of 5.5 favors butyrate production, which was fourfold higher at pH 5.5 than at pH 6.5.[48] Glyphosate, by increasing the gut's pH, reduces butyrate. This study found that certain pathogenic strains of *Bacteroides* increased at higher pH, representing a whopping 78 percent of bacteria present. Crohn's disease, a painful inflammatory bowel condition often diagnosed in people under 35, is linked to an increase in *Bacteroides* and a decrease in colonic butyrate.[49] Symptoms of Crohn's include abdominal pain, severe diarrhea, fatigue, weight loss, and malnutrition. Inflammatory bowel disease is one of the diseases rising in prevalence in the United States, in lockstep with the rise in glyphosate usage on core crops.

Bowels on Bloat

The suffering of a 71-year-old man with a long history of Crohn's disease shows the escalating issues caused by decades-long exposure to glyphosate.[50] This man, the subject of a 2016 case study, was surgically treated for rectal cancer in 1987. He began to develop recurrent bowel obstructions in 2012 that were associated with inflammation in his small intestine, as well as diarrhea and bloating caused by small intestinal bacterial overgrowth (SIBO). SIBO can arise when a shortened or damaged bowel causes impaired nutrient uptake, such that nutrients linger in the gut, which then supports pathogenic bacterial overgrowth.

Throughout 2013 and 2014, the man took repeated courses of antibiotics to keep the bacterial overgrowth in check. Unfortunately, the reduction in bacterial species provides opportunistic space for yeast, such as *Candida*. The man continued to lose weight, so in an attempt to stop the weight loss, he decided to eat more sugar. He started consuming six to eight colas a day, as well as many sugary snacks. Recurrent diarrhea prompted him to resume antibiotic treatments.

The poor man began to experience brain fog and difficulty walking. His wife noticed slurred speech. He fell while showering. But when she rushed him to the hospital, the emergency room doctors couldn't find anything wrong. They sent him home where his symptoms worsened. The next day he returned to the ER. This time, tests revealed that his blood ethanol level was 234 milligrams per deciliter (greater than 80 is considered "driving under the influence," and greater than 300–400 is potentially fatal). Yet he had not consumed any alcohol for over 30 years!

The explanation for this surprising scenario is that the antibiotics were killing off the man's gut bacteria, permitting an overgrowth of yeast, particularly *Candida*, which was further fueled by his high sugar intake. The yeast were then fermenting the sugar to alcohol! Such a phenomenon has a name: endogenous ethanol fermentation, also colorfully known as auto-brewery syndrome. Auto-brewery syndrome causes many uncomfortable symptoms: belching, chronic fatigue, disorientation, dizziness, dry mouth, hangovers, and irritable bowel syndrome. It can also lead to anxiety, depression, and reduced productivity.[51] While this man's case was dramatic, many people today are unknowingly suffering from similar but milder versions of the same syndrome. It's a vicious and debilitating cycle: Inflammation promotes colonization by *Candida*, which delays healing of inflammatory lesions. Studies indicate that 37 to 86 percent of patients with Crohn's disease, ulcerative colitis, and gastric and duodenal ulcers suffer from an overgrowth of *Candida*.[52] We have an epidemic today in candidiasis, which is an overgrowth of *Candida* species in the gut and elsewhere in the body.[53]

As I mentioned in chapter 2, we've also seen a dramatic increase in multiple-cause mortality due to fungal infections: from 1,557 deaths in 1980 to 6,534 deaths in 1997.[54] These deaths are caused by *Candida*, *Aspergillus*, and *Cryptococcus*, and other fungal species. The number of cases of sepsis caused by fungi in the United States increased by 207 percent between 1979 and 2000.[55] Glyphosate is never seen as the main cause of digestive disorders. Instead, doctors tell their patients that their disease is of "unknown origin." But even if the proximate origin is a mystery, the underlying origin seems quite clear: In the war on weeds, our gut microbes are collateral damage.

Blocked Bowels

Along with stomach troubles and digestive difficulties, medical doctors report that painful bowel movements and the inability to defecate are on the rise among both children and adults. In 1989, constipation affected 2 percent of the US population.[56] Today, estimates range from 9 percent to 20 percent.[57] The number of hospital emergency room visits for constipation rose 42 percent over just five years.[58]

What does glyphosate have to do with constipation? The standard American diet is what is usually blamed: Americans don't eat enough fiber, which can make the bowels sluggish. And maybe you've heard the expression, "Sitting is the new smoking." We are also living more sedentary lives than ever before; inactivity can exacerbate constipation. Anyone suffering from constipation is told to drink more fluids, as dehydration can play a role, too. But the herbicide residues in processed and conventionally farmed foods are also to blame. Intestinal paralysis is one of the severe reactions to acute glyphosate exposure.[59] Chronic exposure to glyphosate will cause a similar problem in the digestive tract, though less immediate and less severe. Normally, most of the gut bacteria reside in the large intestine. However, with slow movement of the contents of the gut, abundant nutrients become available to allow microbes to flourish in the small intestine, where they normally don't belong, leading to SIBO.

Disrupted peristalsis leads to constipation, and, on top of that, impaired digestive enzymes cause proteins to be metabolized into ammonia, increasing the pH of the gut. This in turn affects the balance of short-chain fatty acids such as acetate, propionate, and butyrate. Low butyrate starves the colonic mucosal cells, allowing toxic metabolites produced by bacteria and fungi to breach the leaky gut barrier and enter the general circulation.

In order to understand how harmful glyphosate is, we have to look at the individual cells in the human body, and then at the organelles within those cells, and finally at the molecules that pass among the cells. When we examine the biochemistry, the blurred images come into focus. Glyphosate is hurting our guts, preferentially killing the species of bacteria that we need the most. These are the microorganisms that help us with everything from digesting food to synthesizing chemicals that affect learning, memory, and mood. When glyphosate annihilates commensal bacteria, pathogenic bacteria and pathogenic fungi thrive.

Barren Soil

The human gut and the Earth's soil have so much in common. Both depend on diverse species of microorganisms to function properly. The rhizosphere teems with bacteria, fungi, and other organisms too small to see without a microscope. Just like gut microbes, some soil microbes are keenly sensitive to glyphosate. For example, soybeans have a symbiotic relationship with the nitrogen-fixing bacteria *Bradyrhizobium japonicum*. Unfortunately, that nitrogen fixation depends on nickel, which glyphosate chelates, so *B. Japonicum* is unable to fix nitrogen in the presence of glyphosate. When plants can't fix nitrogen effectively, farmers will often add more nitrogen- and phosphate-based chemical fertilizer, running the risk that the chemical fertilizer may then run off into water sources, depleting oxygen in lakes and streams, resulting in eutrophication and leading to toxic algae blooms.

Even plants that have a resistant version of EPSP synthase through GMO technology can suffer from glyphosate exposure. A 2017 study on genetically modified glyphosate-resistant soy plants demonstrated that glyphosate reduced nitrogen-fixing capacity in the root nodules.[60] And glyphosate's toxicity to soil bacteria makes it harder for Roundup Ready plants to take up manganese, even when manganese is plentiful in the soil. The roots of glyphosate-resistant soybean and corn treated with glyphosate can also become heavily colonized by a pathogenic fungus, *Fusarium*, compared to controls.[61]

Disrupting the shikimate pathway in plants can also diminish their capacity to fix carbon and reduce nutrient density. Plants absorb carbon from the atmosphere and use it to build organic compounds.[62] Under normal conditions, as much as 20 percent of organic compounds flow through the shikimate pathway. When plants are under stressful conditions, such as attack from pests or pathogens, or a drought or a heat wave, the amount of carbon that flows through the shikimate pathway increases. Many of the complex molecules that are derived from the shikimate pathway are important for plant defense against stressors.[63] These molecules, such as the polyphenols and flavonoids present in colorful fruits and vegetables, are also important antioxidant defenses that benefit both the plants and the humans who consume them. When we disrupt a plant's ability to make antioxidant defenses for itself, we also disrupt its ability to provide antioxidant defenses to us. Research has shown that glyphosate significantly alters the content of foods derived from exposed plants. Other aspects of plant metabolism are also disrupted by glyphosate. For example, mung beans, a staple in Southeast Asian and Indian cooking, show evidence of disrupted enzymes, DNA damage, and a significant decrease in protein content when treated with glyphosate.[64] Plant photosynthesis is also affected negatively by glyphosate. Roundup Ready soybeans experience yellowing (chlorosis) from glyphosate, caused by impaired photosynthesis because glyphosate disrupts the synthesis of chlorophyll.[65]

Lessons from Karelia

The northern European region of Karelia was a province of Finland until 1939, when the eastern part of it was ceded to Russia. This bitter cold land of dense taiga, large lakes, and ancient volcanoes represents an incredible opportunity for scientists to study lifestyle factors influencing chronic disease. In Finnish Karelia there is a two- to sixfold higher incidence of allergies and a five- to sixfold higher incidence of type 1 diabetes and other autoimmune disorders compared to Russian Karelia.[66] Researchers have long tried to figure out why.

Anyone who has heard of the hygiene hypothesis might think they know the answer. In wealthy countries, we may be inadvertently nurturing autoimmune conditions by being "too clean." Our bodies need exposure to microbes in order to train our immune systems. Water treatment, pasteurization, food sterilization and radiation, antibiotics, vaccines, and decreased exposure to soil—all of these things are associated with "developed" societies and may help explain why Finnish children are less healthy than Russian children.

But there is something else. Finnish children who become diabetic have an overgrowth of *Bacteroides dorei* in their guts.[67] At the same time, only 10 percent of Finnish infants have the beneficial bacteria *Bifidobacteria longum infantis*. Russian babies, by comparison, maintain this important and beneficial strain at high levels throughout infancy.[68] Lack of *Bifidobacteria* and the associated elevated fecal pH promote inflammation-favoring bacteria and gut dysbiosis.[69] Remember that *Bifidobacteria* are vulnerable to glyphosate exposure, and that glyphosate is associated with elevated gut pH.

Finnish children are more exposed to glyphosate than their Russian counterparts on the other side of the border, and they have been for years. The use of glyphosate in agriculture, particularly cereal crops, has increased significantly in Finland since 1999. Since 2001, the Finnish government has subsidized farmers for no-till agricultural practices, in order to minimize phosphate run-off into marine waters.[70] While organic no-till practices can be excellent for ecosystems, glyphosate-based chemical no-till is not.

While Finnish children eat glyphosate-contaminated foods, Russian president Vladimir Putin has been enthusiastic about turning Russia into the organic food capital of the world.[71] Since 2015, Russia has refused to plant GMO crops, and since June of 2016 there has been a near total ban on the use of genetically modified plants in Russian agriculture. It is also illegal to import genetically modified food from abroad. With rich soil that hasn't been ruined by industrial chemicals, Russia has resisted an agrochemical approach to food production. With less exposure to glyphosate in their food, water, and clothing, is it any wonder the children in Russian Karelia are healthier than their Finnish counterparts?

— CHAPTER 4 —

Amino Acid Analogue

Men occasionally stumble over the truth, but most of them pick themselves up and hurry off as if nothing happened.

—WINSTON CHURCHILL

There is an insidious mechanism by which glyphosate is likely causing damage throughout the body, slowly over time. This mechanism explains the alarming rise in various metabolic, neurological, autoimmune, and oncological diseases among exposed humans, and may also account for the extreme distress among many other species, such as monarch butterflies, amphibians, honeybees, and sea coral when they're exposed to glyphosate. If Anthony Samsel and I are right, the consequences are stunning, and we should all be very worried about the future of life on this planet.

Glyphosate is special. Researchers have tried hard, and failed, to find another molecule that similarly suppresses the enzyme EPSP synthase in the shikimate pathway. In fact, Monsanto researchers were surprised to discover that if they replaced the nitrogen atom in glyphosate—remember, glyphosate is $C_3H_8NO_5P$—with oxygen or sulfur, the resulting molecule did nothing to suppress EPSP synthase, despite being almost identical in shape and size to glyphosate.[1]

Researchers have tested over a thousand other molecules that appear to be similar to glyphosate in shape and biophysical properties, but none come close to suppressing EPSP synthase in the shikimate pathway as well as glyphosate.[2] Why did these other chemicals fail? I suspect it is because they are not analogues of the amino acid glycine. As I'll show you, glyphosate can potentially replace glycine in many crucial proteins and shut down their function, and this is what, in my opinion, accounts to a large degree for why glyphosate exposure is contributing to the rise in many diverse diseases.

Protein Primer

Proteins are the workhorses of the body. They perform many different functions: as enzymes that catalyze various reactions; as ion channels that allow minerals such as calcium and potassium to pass through membranes; and as the receptors that respond to signaling molecules such as hormones. Proteins can also be hormones or signaling molecules themselves, such as insulin, which induces glucose uptake by the cells; and enkephalin, a natural endogenous opioid. Other proteins such as collagen provide tensile strength and elasticity to our joints, serving a structural role. Proteins form a big part of hair, nails, skin, and bones.

How are proteins synthesized? They're assembled from amino acids, like beads on a string, following a pattern laid out by DNA. There are only 20 unique coding amino acids (see table A.1 on page 186). Glycine—$C_2H_5NO_2$—is one of them. In fact, glycine is the smallest, simplest amino acid. All amino acids possess a central (or alpha) carbon, an amino group, a carboxyl group, a hydrogen, and an R group. What distinguishes amino acids from each other is the R groups, which are extra pieces, or "side chains," which help amino acids execute the various functions of each protein. Glycine, which has just a single hydrogen atom as its side chain, is often present in situations where a protein requires flexibility, such as an ion channel that needs to open and close like a gate. Glycine, due to its small size, is also useful at a site where the protein binds to a bulky substrate.

Each code from the DNA is made up of a three-letter sequence taken from among the four nucleotide options: adenine (A), guanine (G), cytosine (C), and thymine (T). There are 64 possible ways to arrange the letters (AAA, AAG, AAC, AAT, AGA, AGG, etc.), but only 20 coding amino acids. That means each amino acid has multiple possible codes. All codes for glycine start with two guanines and can be represented as "GGX," where "X" is a "wildcard" that stands for any nucleotide. So GGA, GGG, GGC, and GGT all code for glycine.

Each amino acid has its own specialized enzyme called transfer RNA (tRNA) synthetase that "looks up" each three-letter combination to find which amino acid comes next in the chain. When the DNA code calls for glycine, the glycine-specific tRNA synthetase matches glycine because it has one segment that fits like a key in a lock and another segment that shapes into a pocket where a glycine molecule fits snugly. This is how the cell matches the glycine code to the glycine molecule it needs. The closer the shape matches, the better. If there's no room for a side chain bigger than a hydrogen atom, the tRNA synthetase cannot pick up the wrong amino acid by mistake. But in order to connect the glycine to the preceding amino acid, the glycine's nitrogen atom must be *outside* the enzyme's pocket. It cannot be hiding inside it. And this has important implications for glyphosate.

Glyphosate, like glycine, has only a hydrogen atom as its side chain, so it also fits perfectly into the pocket that's looking for glycine. But glyphosate also has extra material attached to its nitrogen atom. (This extra material is not technically a side chain because it's not attached at the alpha carbon, where side chains attach.) The bulkiness of this extra material may in some cases, depending on the neighboring amino acids, prevent glyphosate from substituting, but it does not prevent glyphosate from fitting into the pocket.

The glycine residues most vulnerable to glyphosate substitution are those with small amino acids to the left, such as glycine itself, alanine, proline, and serine. (Residues are what's left over when two or more amino acids combine into proteins, or smaller chains, called peptides.) Remember that glyphosate kills plants by interrupting the action of the enzyme EPSP synthase. EPSP synthase usually has a glycine residue preceded by an alanine at the site where glyphosate disrupts it. Some organisms have a version of EPSP synthase with alanine instead of glycine, and these organisms are all unaffected by glyphosate. Many of the pathogenic bacteria and fungi that flourish in the gut are protected from glyphosate because of the replacement of glycine with alanine. Other organisms have evolved to protect themselves by exchanging a neighboring amino acid for a bulkier one, crowding out the glyphosate molecule and interfering with its substitution for glycine; these organisms are also insensitive to glyphosate.

Directed Evolution?

Statistical studies on genetic mutations in humans have revealed that certain amino acids are more vulnerable to a change in the DNA code that causes it to get replaced by a different amino acid (aka substitution error) than others.

The amino acid arginine is the most vulnerable to substitution. The DNA codes for arginine are CGT, CGC, CGA, CGG, and AGA. You may notice that four of these contain the nucleotide pair "CG." The CG pair mutates more readily than other codes. In fact, CG → TG or CG → CA substitutions are at least 13 times more likely to occur than any other substitution. This explains why arginine is more vulnerable to substitution error than other amino acids: Four out of its five codes contain the CG sequence.[3]

After arginine, glycine is the second most commonly mutated amino acid. We also know that these mutations occur most often when there is another glycine to the left of the mutated glycine: GG sequences.[4] But researchers aren't sure why glycine mutates at such a high rate. They also don't understand why a glycine with glycine as its left neighbor is especially vulnerable. I propose that the second glycine in a GG sequence is an ideal spot for glyphosate to swap in. As the smallest, simplest amino acid with only a single hydrogen atom as its side chain, glycine leaves the most room "next door" for the added bulk of glyphosate's extra methylphosphonyl group, which is attached to its nitrogen atom.

This phenomenon has implications far beyond glyphosate. It suggests there is a biological mechanism in place to detect stressors on certain peptide sequences and to impose pressure to alter them. Disease-causing genetic mutations may actually be indicators of areas of stress. Exactly how an organism could increase mutational pressure on a specific stressed location in a peptide sequence is unclear, but it seems likely it is happening. Why it's happening is an equally important question.

While students are commonly taught that genetic mutations leading to evolutionary change are random, complete randomness makes little biological or evolutionary sense. If biological mechanisms could detect which specific proteins, and which specific regions within those proteins, are the most stressed, evolutionary strategies targeting those stressed regions would be an efficient route for adaptation to environmental change. This hypothetical mechanism might not be "smart" enough to select the best changes, in which case some mutations would be bad choices and the modified organism would fail to thrive. Classic Darwinian evolution, then, would play a role in weeding out such failed modifications. My hypothesis is that, because glyphosate is disrupting protein function by swapping in for glycine at critical places, it is creating stress in the region around those specific susceptible glycine residues, causing an increased mutation rate at those spots.

You may have learned about Lamarckian evolution in biology class. Jean-Baptiste Lamarck was a French naturalist, born in 1744, who proposed that acquired traits could be passed on to future generations. Lamarck understood that life is not fixed. When the environment changes, organisms have to change in order to survive. The theory of Lamarckian evolution remains controversial and Lamarck himself tends to get a bad rap, but he may have been onto something important. He was suggesting a phenomenon I think of as "directed evolution," whereby a living organism is able to recognize where environmental stressors are causing the most damage, and then tweaks those sections of the genome to try to find a solution. Although most altered proteins will likely fail, sometimes there are wins, and those organisms flourish. This is the principle of Darwin's idea of survival of the fittest.

Interestingly, the human protein with the greatest number of disease-causing substitution errors—called single nucleotide polymorphisms, or SNPs, and pronounced "snips"—is glucose-6-phosphate dehydrogenase (G6PD). G6PD is a protein that is active in red blood cells, and it is especially susceptible to glyphosate substitution, as I will explain more fully in chapter 5. If glyphosate continues to be a part of our environment, "successful" organisms will be those that acquire mutations within their genomes that lose vulnerable glycine residues and produce proteins that can still function properly.

Other Disease-Causing Amino Acid Analogues

Protein synthesis is an error-prone process: Scientists estimate that about 15 percent of all the proteins in the body have at least one mistranslated amino acid.[5] Cells are careless during protein assembly. Efficiency is privileged over accuracy and mistakes are weeded out later during a complex folding process that shapes the protein in the right way (or not) to do its assigned tasks.

Sophisticated biological mechanisms determine whether a protein has folded correctly, and whether it's able to do its job. If it is defective, specialized protein molecules called ubiquitin bind to the protein, tagging it the way a forest manager might mark a diseased tree for later removal, as if to say "clear this protein." Other proteins that specialize in disassembly break down the tagged protein, and the assembly process begins again, aiming for a better product this time around.

The idea that amino acid analogues can substitute for coding amino acids during protein synthesis and can subsequently cause disease is not disputed. Several amino acids have natural analogues that are known to be toxic. Let me give you a few examples.

Dysfunction of myelin basic protein is linked to multiple sclerosis, a serious disease of the brain and spinal cord. Multiple sclerosis affects the myelin sheaths that cover your nerve fibers to protect nerve transmission. Proline, an amino acid critical to myelin basic protein, has a natural analogue produced in sugar beets.[6] This is why, some scientists suggest, multiple sclerosis is more prevalent in people living close to sugar beet farms.[7]

Fine fescue is a hardy turfgrass that grows year round, often under difficult conditions. Several species of fine fescue release large amounts of a noncoding amino acid analogue of phenylalanine, known as meta-tyrosine, into the rhizosphere. Phenylalanine induces root growth. Meta-tyrosine, on the other hand, is toxic to cells, making it a natural herbicide that inhibits root growth.[8]

Glufosinate is a naturally occurring amino acid analogue of glutamate. It is increasingly used as an herbicide in agriculture due to the appearance of glyphosate-resistant weeds.[9]

And an analogue of serine, BMAA, is produced by cyanobacteria. BMAA is believed to cause an ALS-like disease that was an epidemic in Guam in the years following World War II. It resulted in protein folding errors that eventually resulted in disease, sometimes decades later.[10] BMAA produced by cyanobacteria is also believed to be responsible for the amyloid beta plaque brain pathology found in stranded dolphins, as mentioned in chapter 2.

One more: l-canavanine, an amino acid analogue of l-arginine, is believed to be responsible for the death of Chris McCandless, the protagonist of Jon Krakauer's book *Into the Wild*.[11] L-canavanine naturally occurs in the wild potato seeds that McCandless consumed in large quantities in the months preceding his death.

We also know that even coding amino acids can be toxic in excess, due to substitution for other coding amino acids. Glycine itself can be problematic because it can substitute for alanine, the main mechanism of toxicity in excessive glycine supplementation. High levels of homocysteine, another amino acid, are associated with increased risk of cardiovascular disease. Homocysteine substitutes for methionine when methionine is deficient. In fact, this is so common that a mechanism is in place to yank the offending homocysteine back out of the chain. This mechanism involves a conversion to homocysteine thiolactone, an inflammatory agent and one of the strongest predictors of cardiovascular disease.[12]

Evidence from the Shikimate Pathway

The most stunning evidence that glyphosate substitutes for glycine during protein synthesis comes from the effects glyphosate has on EPSP synthase, the enzyme in the shikimate pathway that glyphosate disrupts, which is attributed as a primary mechanism behind its ability to kill weeds.[13] Multiple studies have been published on this topic, although scientists have been unsure of the mechanism of its toxicity and are unable to adequately explain what they were witnessing. A currently accepted theory of glyphosate's disruptive mechanism is that it competes with phosphoenolpyruvate (PEP), a substrate that normally fits snugly into a specific pocket of EPSP synthase. In other words, glyphosate "pretends" to be the PEP substrate, which prevents PEP from binding to EPSP synthase.

As you know, different species of plants and microbes have different versions of the enzyme EPSP synthase, but many of them share a particular peptide sequence: LGNAG. The EPSP synthases of petunia, *E. coli*, tomato, *Arabidopsis*, soybean, and corn all have this sequence. Notice that two Gs are in this peptide sequence. "G" stands for glycine. The second glycine is remarkable: Multiple species of plants and microbes have a version of EPSP synthase where glycine has been replaced by alanine. These versions are invariably associated with resistance to glyphosate.[14]

A team of scientists, led by Monsanto chemist and coinventor of Roundup Ready technology, Stephen R. Padgette, PhD, was able to modify versions of EPSP synthase in the lab by replacing this glycine residue with alanine in their experiments.[15] They used a method called site-directed mutagenesis to change the DNA to code for alanine instead of glycine. All the original versions they tested were sensitive to glyphosate. All the alanine-substituted variants were resistant to it.[16]

These lab-created variants of the enzyme with glycine swapped out for alanine do not bind as well to the PEP substrate because alanine has an extra methyl group (CH_3). This methyl group crowds the enclosure formed by the protein where PEP needs to fit snugly. A different EPSP modification that swaps out two nearby amino acids but not the second glycine is also strongly resistant to glyphosate, but it also does not disrupt PEP binding, an ideal situation. One of the two substitutions involves replacing the amino acid threonine with the bulkier amino acid isoleucine, which has an extra methyl group. This methyl group likely crowds out the glyphosate next door, preventing it from substituting for glycine during protein synthesis. It would also likely interfere with PEP binding if it weren't for a second substitution, where proline is swapped out for serine. This second substitution introduces extra space in the pocket such that PEP can now fit comfortably, albeit in a pocket that has shifted slightly from its original position.[17]

Curiously, when alanine is substituted for the second glycine in the LGNAG peptide sequence, multiple other inhibitors that work by substituting for the substrate PEP are just as effective as they were with glycine in that place. These other inhibitors should still be able to inhibit PEP binding, because if PEP fits, they should, too. So the big question is, why does glyphosate get stymied by this modification when other inhibitors don't? The answer is that glyphosate works through an entirely different mechanism than the others do: It substitutes for glycine within the enzyme's peptide sequence rather than substituting for PEP as a stand-alone molecule. Remember that the commonly accepted explanation for glyphosate's mechanism of toxicity is that it "pretends" to be PEP, which prevents PEP from binding to EPSP synthase. It's an understandable assumption because multiple other inhibitors work in precisely this way, but there is another difference. Glyphosate doesn't inhibit PEP binding to many other enzymes that use PEP as a substrate.[18] This is likely because these enzymes don't have a glycine residue at the active site that can be easily substituted by glyphosate. (The active site is the location on the protein where the reaction takes place.)

Manipulating Maize

CRISPR, or Clustered Regularly Interspaced Short Palindromic Repeats, is a powerful new gene editing technology that allows scientists to easily and precisely edit the DNA of any genome. For the development of CRISPR, Drs. Emmanuelle Charpentier and Jennifer Doudna were awarded the 2020 Nobel Prize in Chemistry.

Potential applications of CRISPR are many and various. They include correcting genetic defects, treating and preventing the spread of disease, and improving crop resistance to herbicides, drought, and other stressors. CRISPR is a powerful tool, but we don't yet understand exactly how it works, or what unintentional consequences it might provoke in a cell. Many bioethical issues are also associated with its use.[19]

In 2018, researchers from DowDuPont applied CRISPR to develop crop resistance to glyphosate.[20] The DowDuPont team of 12 researchers first examined the genetic sequence of several different versions of EPSP synthase drawn from a variety of glyphosate resistant weeds and microbes. They observed that the second glycine in the LGNAG peptide sequence plays a crucial role in glyphosate susceptibility. Namely, anytime that second glycine is replaced with something else, the enzyme is protected. The researchers then made tweaks to the DNA code to produce seven different EPSP variants of maize. In all seven variants they started by substituting alanine for the second glycine in the sequence.

To fully succeed, the team had to make other modifications to the enzyme in order to increase the size of the PEP binding site, which had shrunk because of alanine's extra methyl group. They found various ways to accomplish this, modeled after successful glyphosate-resistant mutant weeds. What this means in terms of seed development is disturbing: Chemical companies can use CRISPR technology to create designer variants of the plant's original EPSP synthase in order to produce patentable glyphosate-resistant seeds that will not be considered GMO crops by regulators (because there was no insertion of a gene from a foreign species into the plant's genome). Unless we do something to prevent this, we can expect to see even more glyphosate used on our food crops in the future.

Presciently, the DowDupont scientists explained: "All known mutations function directly or indirectly at the contact between G101 [the second G in GNAG, equivalent to residue 96 in *E. coli*] and a phosphonate oxygen of glyphosate."[21] (Because the two proteins are not exactly the same length, the version of the protein in the plant has the critical glycine residue displaced by five residues compared to *E. coli*, from position 96 to 101.) These scientists were not correct, but they were close. They refer to the "contact between" the glycine residue at 101 and the phosphonate of glyphosate. However, the phosphonate of glyphosate is actually *part of* the "glycine" molecule: glycine has been swapped out for glyphosate. All of these mutations are crowding out glyphosate at the time of protein assembly.

The DowDuPont scientists were especially excited about an EPSP variant, called D2c-A5, that involved several modifications around the active site for PEP, including a glycine-to-alanine substitution. This variant was effective because it bound well to PEP but was completely insensitive to glyphosate. They then did a kind of "inverse" experiment where they modified D2c-A5 in only one way: restoring glycine where the alanine substitution had been. Restoring glycine resulted in a near complete restoration of glyphosate sensitivity. It also somewhat decreased the enzyme's catalytic activity, and the scientists proposed that this was because PEP's binding site was now too large, a consequence of losing alanine's methyl group.[22] This experiment clearly shows that glyphosate exerts its effects through a mechanism other than displacing PEP in the pocket.

A version of EPSP synthase where leucine, an essential amino acid, replaces proline at residue 101 was discovered by scientists in 2005. These researchers found it hard to explain how this alteration worked to suppress glyphosate, because it did not really disrupt the site where PEP binds to EPSP synthase. However, the researchers realized that leucine, unlike the original proline residue, crowded the glycine at residue 96. They suggested that somehow that might impact glyphosate more than PEP, without appreciating that the real reason is that glyphosate displaces glycine, not PEP.[23] Ominously, this version of EPSP synthase is the natural one produced by the pathogen *Staphylococcus aureus* (MRSA), which, as mentioned before, has become a major problem in hospitals due to its multiple antibiotic resistance. This staph bacteria's natural resistance to glyphosate affords it an advantage that allows it to flourish in the presence of glyphosate that kills off other bacteria.

Incorporated into Our Protein

We have been led to believe that glyphosate passes through the body quickly and is rapidly excreted in feces and urine, without accumulating in our tissues.

But Monsanto's own studies prove otherwise. Bluegill sunfish are spirited and strong. A popular game fish, they fight like crazy when a fisherman hooks them. In 1989, Monsanto researchers exposed bluegill sunfish to glyphosate and then used carbon radiolabeling to identify how much was present in the animals' tissues. At first, only 17 to 20 percent of the radiolabel could be identified as glyphosate. After the researchers used a digestive enzyme to break down the proteins into amino acids, 70 percent of the radiolabel could be tagged as glyphosate. The researchers concluded that the glyphosate was initially hidden because it "was tightly associated with or *incorporated into protein*"[24] [emphasis added].

This conclusion should send a chill down your spine. The glyphosate had been taken into the tissue of the bluegill and *become part of it*. The implication is that glyphosate *becomes part of* the tissues of all the species of animals and plants that are exposed to it. Although the incriminating results of this research was never made available to the public, my colleague, Anthony Samsel, obtained a copy of the unpublished study from the Environmental Protection Agency via a Freedom of Information Act request.[25]

DuPont researchers have also conducted experiments with radiolabeled glyphosate involving chickens and goats.[26] Scientists found glyphosate in tissue samples of both animals, as well as in the chickens' eggs. As with the bluegills, much of the glyphosate was at first hidden because it was incorporated into the animals' own proteins at the molecular level. It showed up only after the proteins had been broken down into their component amino acids, freeing up glyphosate as a separate molecule that could now be detected by the technology. Glyphosate wasn't a free molecule circulating in the tissue. It was an integral part of "Frankenstein" proteins that had replaced the animals' own.

Disrupted Protein Synthesis

I'm a cell and I'm busy managing my life. A lot of my life involves making proteins. I use them for everything, including receptors, enzymes, ion transporters, and structural support. I also make proteins that make other proteins, and proteins whose job it is to bust open faulty proteins. Once a new protein is made it gets folded. This puts the protein into the correct shape it needs to do its job.

But when glyphosate, or any other amino acid analogue, starts damaging the proteins I need, I get in trouble. As I've explained, protein synthesis is a sloppy process. Proteins making other proteins are busy grabbing codes and using them to assemble a long string of "paper dolls." But the proteins synthesizing other proteins are in a hurry. They don't pay attention to the details. Think of it as a fast and busy assembly line where everyone is in a rush. They can easily grab glyphosate when they see the code for glycine, and they slip it into the paper doll string by mistake.

But living beings have a way to fix these biochemical mistakes. There are other proteins whose function is to notice when something goes wrong. These proteins are like quality control managers on an assembly line. They put a marker on the faulty protein to show it's been rejected, signaling to other proteins to pull apart the damaged goods into component amino acids. The process of protein synthesis and protein dissolution is going on all the time in all living beings. However, with the presence of glyphosate, you have a lot more proteins going into the reject pile. This creates a useless and damaging cycle because so many proteins are broken.

Cells have an instruction code, an RNA sequence, that tells them how to manufacture a particular protein. RNA sequence analysis, a method that first appeared in 2008, allows scientists to look collectively at protein expression. It shows clearly how glyphosate is substituting for glycine during protein synthesis. In 2016, a team of plant pathology and environmental scientists analyzed the genomes of the microbes living in the root zone of plants. They discovered many disturbances in gene expression due to glyphosate exposure. Among many other observed alterations, there was a big increase in synthesis of proteins that are needed to synthesize more proteins. And there was also an increase in proteins that break proteins apart into component amino acids.[27]

Proteins involved in both protein synthesis and protein degradation are disrupted in response to glyphosate. Why would this be? Because, when glyphosate gets incorporated into proteins, it often makes them dysfunctional, in plants as well as in animals, and presumably also in humans. In order to replace the defective proteins, the living organism must first disassemble the faulty proteins and then synthesize fresh versions from the liberated component amino acids in a renewed attempt to get it right.

Scientific Pushback against an Inconvenient Truth

Some biochemists reject the idea that glyphosate substitutes for glycine during protein synthesis. In 2019, a team of London-based researchers claimed to prove that glyphosate does no such thing. These scientists exposed human breast cancer cells to glyphosate for 6 days. They then used a sophisticated technique called tandem mass tag labeling to identify short chains of amino acids purportedly containing anomalously heavy glycine molecules. The proteins were put through a mass spectrometry analysis to detect these unusually heavy proteins.[28]

The scientists tested the samples for two distinct modifications: glyoxylate-modified cysteine as well as glyphosate substitution for glycine. They were looking for glyoxylate modification, because they hypothesized that glyphosate might get broken down to glyoxylate, which is capable of binding to cysteine residues. No glyoxylate-modified cysteines were detected in either the control or the treatment group. In contrast, the researchers found a substantial signal for glyphosate presence in several short peptides in the treated samples . . . but they also found an equally strong signal in the untreated samples!

There were several missteps with this experiment, problems so pronounced that, in my view, they invalidate the study's conclusions. Both the exposed and unexposed breast cancer cells were maintained on an enriched nutritional formulation, Dulbecco's Modified Eagle Medium, which contains amino acids and vitamins and has a high concentration of glucose. The medium itself may have been contaminated with glyphosate. Since the scientists didn't test it, there was no way to tell. Another misstep: Cells were supplied with blood drawn from fetal bovine serum, also untested for glyphosate contamination. Glyphosate has been found in cows' organs and tissues and has been detected in human umbilical cord blood, as well.[29] Finally, the experiments were conducted on a highly aggressive human breast cancer cell line called MDA-MB-231, derived from a tumor in a 51-year-old woman in the 1970s.[30] The cells have been maintained in culture since they were harvested, and they have probably been chronically exposed to glyphosate for decades. They could have accumulated substantial numbers of misfolded glyphosate-contaminated proteins before the experiment even began, which could have then been recycled . . . into newly contaminated proteins.

So when these scientists stated that "the data conclusively shows that all candidate substituted peptides are false discoveries," I would argue they are wrong.[31] It's more likely that *all* candidate substituted peptides were *true* discoveries, because both the treatment *and* the control group had been exposed to glyphosate prior to and during the experiment, allowing both the treatment group and control group to accumulate misfolded glyphosate-contaminated proteins in near equal amounts.

It's worth noting that 10 out of the 15 short peptides had a small amino acid immediately to the left of the substituted glyphosate molecule. As I have mentioned, this offers an ideal opportunity for glyphosate to fit its extra methylphosphonyl group into the protein structure. For context, 10 out of 15 is a higher frequency than you would expect in a random sample. Even if there weren't other issues with the experiment, it does not seem plausible that *all* the detections were false alarms, especially since the methodology did not detect any false alarms for the other substitution the researchers were trying to detect: the glyoxylated cysteines.

Deadly Consequences

The consequences of glyphosate's ability to incorporate itself into proteins in place of the coding amino acid glycine are far-reaching and potentially catastrophic. Nearly every protein of any length has at least one glycine residue. There is a large set of proteins for which at least one specific glycine residue is essential for proper function. Researchers can determine which amino acids in a given enzyme class are important by aligning multiple protein sequences with a shared evolutionary relationship taken from many different species. Sometimes these alignments have sequences or individual amino acids that are identical across a large number of species, even across phyla—from amoebas to cabbages to chickens. When the same amino acid shows up in these alignments across multiple versions of a particular protein, it is considered to be "highly conserved," and this almost always means that the amino acid serves an important function in the protein. For example, the muscle contractile protein myosin has at least a dozen glycine residues that show up consistently in many different versions.[32] The human version of the enzyme has a highly conserved glycine residue at location 699. If this is replaced by alanine, the protein's ability to contract is reduced to 1 percent of its capacity.[33]

Collagen is the most common protein in the body, making up 25 percent of the body's proteins. It has a unique GxyGxyGxyGxy pattern over large swaths of the molecule. In this pattern, every third residue is a glycine (you'll remember that "G" stands for glycine and the lowercase "x" is a wildcard, meaning that any amino acid could go there, including glycine). This pattern is essential for collagen to fold properly into its elegant triple-helix structure. It is this structure that gives collagen its unique properties of tensile strength, elasticity, and ability to hold water. Mutations in several different glycine residues in collagen weaken connective tissues, resulting in a condition known as Ehlers-Danlos syndrome. People who have Ehlers-Danlos syndrome often suffer from loose joints, stretchy skin, and abnormal scar formation.[34]

While Ehlers-Danlos syndrome is a genetic disorder, glyphosate substituting for glycine residues within collagen is an environmentally induced disruption that may be contributing to the widespread back, shoulder, neck, knee, foot, and hip pain so many of us experience today. This pain often necessitates surgery to repair injured joints. Among people in their 70s, the prevalence of knee surgery among women has increased nearly 11-fold, and the prevalence of hip surgery has increased 9-fold, over the past two decades. Among men, the increase has been even more dramatic, with a 26-fold increase in prevalence of hip replacement surgery and a 15-fold increase in knee replacement surgery.[35] Joint replacement is expected to become the most common elective surgical procedure in coming decades. There are other reasons for joint damage and attendant surgeries, including stress on the body caused by obesity, improved surgical techniques with better outcomes, and the profitability of surgical interventions. But it is also likely that glyphosate's disruption of collagen increases the injury, necessitating corrective action.

Many ion transport proteins depend on glycine in a critical hinge region of the protein, which is embedded in the plasma membrane of the cell. This hinge is essential for the opening and closing of the protein to control ion flow across membranes. There is also an interesting class of proteins that have a transmembrane segment containing a "glycine zipper motif" where glycine is regularly spaced by three intervening amino acids. (In chemistry, a "motif" is a pattern of amino acids in a protein sequence.) The pattern is GxxxGxxxGxxxG for amyloid beta, the protein whose misfolding and precipitation into plaque is linked to Alzheimer's disease. This same GxxxG pattern is also associated with prion proteins that cause prion diseases such as mad cow disease in bovines, chronic wasting disease in deer, and Creutzfeldt-Jakob disease in humans.[36]

We all probably know people suffering from Alzheimer's disease, which is a chronic degenerative condition and the most common form of dementia. At least 5 million Americans currently have Alzheimer's and an estimated 14 million Americans are projected to have the disease by 2060.[37] Alzheimer's affects the parts of the brain that control working knowledge, language, memory, and thoughts. As their condition deteriorates, Alzheimer's patients can experience so much cognitive confusion that they become unable even to feed themselves. They might pick up a fork and just look at it, unsure what it is or how to use it. With their brain cells wasting away, they lose the common sense that most of us take for granted.

Studies have shown that amyloid beta depends on the glycine zipper motif to maintain its alpha helix structure that penetrates the cell membrane. If that glycine is replaced with other amino acids, the amyloid beta forms soluble beta sheets instead. These beta sheets stay in the cytoplasm and eventually accumulate, glomming together to form fibrils that become the characteristic Alzheimer's plaque.[38] This process of misfolding into beta sheets is called amyloidosis, and we also see it happening with other proteins associated with other pathologies.[39]

It is not a coincidence that the prevalence of chronic diseases such as Alzheimer's have been rising in step with exposure to glyphosate. While we can't blame every aspect of an illness on this pernicious chemical, we can surmise that it is contributing to the downturn in human health by disrupting our proteins. Dozens, if not hundreds, of proteins may be severely disrupted by glyphosate substitution for glycine. Many receptors depend upon glycine residues to attach to the cell membrane and receive a signaling molecule, such as insulin or low-density lipoprotein (LDL), which delivers fat molecules to the cells.[40] You've no doubt heard LDL referred to as bad cholesterol, because it can sometimes collect in the walls of blood vessels and it's associated with an increased risk of heart attacks and strokes. When patients have lab tests that reveal high levels of LDL in their blood, medical doctors prescribe statins. Statin drugs are linked to a long list of worrisome side effects.[41] Glyphosate substitution in the insulin or LDL receptors could be a factor in both insulin resistance (type 2 diabetes) and elevated serum LDL associated with cardiovascular disease.

Once we understand how glyphosate disrupts the shikimate pathway by substituting for glycine and disrupting or even destroying hijacked proteins, we have a model to understand how glyphosate can disrupt many other proteins by acting similarly as a biochemical imposter. In the next chapter, we'll look specifically at proteins that bind molecules that contain phosphate at a site with at least one highly conserved glycine residue. EPSP synthase exemplifies this scenario. The implications of glyphosate substitution in this broad class of proteins are terrifying.

— CHAPTER 5 —

The Phosphate Puzzle

As a society we are like a freight train heading for a broken bridge.

—DIETRICH KLINGHARDT, MD, PhD

Does glyphosate have a toxic mechanism that damages a broad spectrum of proteins? And, if so, can this explain the long list of chronic diseases whose prevalence has been going up in line with the rise in glyphosate usage on core crops? Yes. And yes. I've been analyzing the scientific literature specifically on glycine residues in proteins for over five years. What I now believe is that glyphosate damages a large number of proteins by exerting a specific toxic effect that disrupts protein function. And that there's a broad class of essential enzymes that share a common feature—they bind phosphate—that causes them to be very sensitive to glyphosate. It's the phosphate-binding capacity that makes these enzymes especially vulnerable. At the same time, this capacity is essential for their function, and these proteins serve diverse needs in the body.

As the smallest amino acid, glycine plays an important role in many proteins, especially by creating space that welcomes the reception of a ligand. (A ligand is a molecule that binds to an enzyme and serves as a substrate to the enzyme's reaction.) By substituting for glycine, glyphosate may deny that space to the ligand by filling it with its bulky, negatively charged methylphosphonyl tail. Think of it as glyphosate stealing the ligand's seat in a busy cafeteria, or as a game of Duck Duck Goose where glyphosate slips into the empty spot.

The substitution is especially pernicious when the ligand is a negatively charged essential small molecule containing phosphate. Glyphosate's negative charge and the phosphate's negative charge (PO_4^{-2}) will repel each other the way two magnets do when you try to push them together with their dipoles in opposition. This repulsion, combined with glyphosate occupying the space where phosphate is supposed to fit snugly, can wreck the protein's ability to do its job.

There are many small molecules containing at least one phosphate anion that play multiple essential roles in biology. One you might be familiar with is adenosine triphosphate (ATP), which has a tail of three phosphate anions hooked together. ATP is considered the "energy currency" of cells. Many proteins derive the energy they need to perform their functions by breaking off the last phosphate from ATP. Another molecule you are now familiar with that contains a phosphate anion is phosphoenolpyruvate (PEP), the molecule that the enzyme EPSP synthase binds to in the shikimate pathway.

Other phosphate-containing molecules that participate in important ways in enzymatic activities and cellular signaling include guanosine triphosphate (GTP); nicotinamide adenine dinucleotide (NAD), which is made from niacin, a.k.a. vitamin B_3; pyridoxal-5'-phosphate (PLP), which is made from pyridoxine, a.k.a. vitamin B_6; flavin adenine dinucleotide (FAD) and flavin mononucleotide (FMN), which are made from riboflavin, a.k.a. vitamin B_2; and glucose-6-phosphate (G6P). As you can see, many of these are essential B vitamins. They play a crucial rôle in helping a living organism with its metabolism.

Several experiments have shown that glyphosate can suppress various enzymes that bind molecules containing phosphate. In addition to EPSP synthase, these include ribulose-1,5-bisphosphate carboxylase/oxygenase (RuBisCo), an enzyme in plants that is essential for photosynthesis and the most common protein known to biology.[1] In human cells, glyphosate suppresses succinate dehydrogenase in the citric acid cycle. Succinate dehydrogenase is an essential enzyme for metabolizing glucose and generating ATP. Glyphosate also suppresses glucose-6-phosphate dehydrogenase (G6PD), an enzyme present in large amounts in red blood cells that plays an important role in antioxidant defenses.[2]

I have a theory that can predict whether a protein will be sensitive to glyphosate. Proteins most at risk have what I call a glyphosate susceptibility motif. (As a reminder, a "motif" in chemistry is a pattern of amino acids in a protein sequence.) This glyphosate susceptibility motif is characterized by three basic principles:

1. The protein binds phosphate at a site where at least one glycine residue is highly conserved and essential.
2. At least one glycine residue at this site has a small amino acid immediately to the left.
3. There is at least one, and preferably two or more, positively charged amino acids nearby that can securely hold glyphosate's methylphosphonate group in place.

Remember that glyphosate disrupts the enzyme EPSP synthase in the shikimate pathway by interfering with binding of the phosphate group attached to phosphoenolpyruvate (PEP), as we discussed in chapter 4. Almost any protein that involves binding to a phosphate-containing molecule has at least one, and usually more than one, highly conserved glycine residue at the site where phosphate binds.

Wasting Energy

Phosphate plays an important role in living organisms. ATP has three phosphates linked together and attached to its adenosine molecule: adenosine-phosphate-phosphate-phosphate. Cells produce ATP mainly by metabolizing sugar in mitochondria during the citric acid cycle. To convert ATP to ADP, the last phosphate is removed. That conversion releases the energy that was tied up in the phosphate bond, which can then be used to perform useful work.

When a protein folds, it creates a specific three-dimensional pocket to snugly fit each of the ligands that the protein binds to—as a reminder, ligands are molecules that bind to enzymes and serve as substrates to the enzymes' reactions. The protein also often configures a complementary charge distribution around the ligand to act like a magnet to secure the ligand in place. So, ATP is a negatively charged ligand for all enzymes that use it as an energy source.

I know this is technical but stay with me. An analysis of 168 proteins that bind ATP found that glycine, as well as the three positively charged amino acids—arginine, lysine, and histidine—were more likely to be found at an ATP-interacting site where the conversion from ATP to ADP takes place. Positively charged amino acids can bond to the negatively charged phosphate unit to secure it in place. Glycine, being the smallest amino acid, provides flexibility as well as room for the bulky phosphate anion.[3]

This model fits well for EPSP synthase. *E. coli*'s EPSP synthase has two lysine residues, as well as a histidine and an arginine residue, that are present at the PEP binding site, in addition to the glycine residue that gives it vulnerability to glyphosate toxicity.[4] Protein folding is a complex, nearly miraculous process that takes place in real time as the protein is assembled, left to right, from its individual amino acids, according to the DNA code. It is likely that, when glyphosate assembles into EPSP synthase at the spot where glycine usually goes, its methylphosphonyl group is secured into the pocket intended for phosphate, aided by electrostatic bonding to the neighboring positively charged amino acids as they fold into place in parallel. The consequence, which would hold true for any protein that normally binds phosphate, is that the protein's ability to utilize phosphate-containing molecules in its reactions is disrupted.

For example, the muscle contractile protein myosin binds to ATP, detaches the last phosphate, and uses the released energy to effect muscle contraction. The site where myosin binds to ATP contains a glycine-rich motif, referred to as the phosphate-binding loop (P-loop) motif, with at least two, and sometimes three, glycine residues.

Myosin has many highly conserved glycine residues, but the center one in its ATP binding site is crucial for its ability to translate the energy in the ATP molecule into useful motion. At least 80 out of 82 known myosin versions in different species have been found to have GESGAGKT as their instantiation of the P-loop motif (GxxGxGKT) at the ATP binding site.[5] Mutations in the middle glycine have particularly devastating effects on myosin function. This glycine plays two roles. One role is to make room for the bulky phosphate anion. The other role is to encourage the phosphate ion to split from ATP. A mutation where this glycine is replaced by alanine (a very small change) reduces the protein's ability to detach the phosphate and therefore to effect muscle movement.[6] Glyphosate replacement could be equally if not even more catastrophic.

In an important study published in 2013, biochemists from Ohio and Indiana conducted a detailed analysis of proteins that were either upregulated or downregulated when *E. coli* bacteria were exposed to glyphosate.[7] At least 10 ATP-binding sites for various protein complexes were found to be upregulated. Upregulation means an increase in the synthesis of the protein, which can happen when the protein is not working properly. Think of it this way: If you hire incompetent workers for a job, you have to hire more workers to get the job done. And even then the job might not be done correctly. Glyphosate substitution for one or more of the critical glycines at the ATP binding site creates a need for increased production of all these inefficient ATP-binding proteins. Many of the disrupted molecules are transport proteins. They use the energy in ATP to support the transport of critical nutrients across the membrane and into the cell. (See table A.2, page 187, for the specific ATP-binding proteins disrupted in *E. coli*.)

Epigenetic Effects through Phosphorylation

Kinases are a large class of enzymes that typically transfer the terminal phosphate from an ATP molecule to an amino acid within a protein.[8] Serine, threonine, and tyrosine are the amino acids that can be phosphorylated in this manner. Phosphorylation (that is, adding phosphates) is one of many epigenetic transformations that cause proteins to behave dramatically differently from the way they would behave otherwise. Most enzymes that are affected are activated through phosphorylation, but some can be suppressed by phosphorylation. Many have multiple sites where phosphates can be added, some of which activate the protein and some of which suppress it, in a sophisticated and complex control mechanism.

Phosphorylation can cause a protein to become more soluble. When this happens, a protein might move from the cell membrane into the cytoplasm. These changes lead to complex outcomes as kinases start attaching phosphates to other kinases, which activates a cascade of profuse activity, and can ultimately effect specific outcomes that change cellular behavior in dramatic ways. The overwhelming and inappropriate immune response that sometimes occurs in patients suffering from an infectious disease like COVID-19 provides a good example. Signaling mechanisms in response to the coronavirus result in a phosphorylation cascade that ultimately causes immune cells to release signaling proteins called cytokines. A sudden and large release of cytokines is a cytokine storm—an out-of-control inflammatory response that can cause widespread tissue damage. A cytokine storm brought on by COVID-19 can be deadly. Cytokine storms are a consequence of immune deficiency, which, as we will see in chapter 10, can be caused or exacerbated by chronic glyphosate exposure.

There is a highly conserved glycine-rich motif at the ATP binding site of kinases, which has a characteristic GxGxxG pattern.[9] The middle G of this pattern is conserved in 99 percent of protein kinases. Based on the kinetics of kinases at the site where ATP binds, glyphosate substitution for either of the first two glycines in the GxGxxG motif commonly found in kinases will have significant effects.[10]

We know that substituting the amino acid aspartate for the middle glycine in a kinase in corn results in a substantially *increased* ability to yank phosphate off of ATP.[11] Aspartate, like glyphosate, is negatively charged and bulkier than glycine. It is a good model for how glyphosate might be expected to behave. A glyphosate-substituted "mutant" will be much more effective at yanking the phosphate off the ATP molecule but less able to grab it and attach it to the phosphorylated protein. Phosphate's detachment from ATP is the rate-limiting step of the reaction, so doing this more effectively means achieving a greater rate of phosphorylation. However, with glyphosate or aspartate in the way, there is also a greater likelihood of metaphorically tossing the phosphate anion on the floor rather than attaching it to the substrate. This essentially wastes the energy in the ATP phosphate bond, putting more pressure on the mitochondria to produce adequate amounts of ATP.

Many studies relate diseases to abnormal phosphorylation of various proteins, such as hyperphosphorylated tau proteins in Alzheimer's disease and disturbed phosphorylation in cancer.[12] The B-Raf kinase is a major regulator of signaling pathways that control cellular division and differentiation. The oncogene BRAF, which codes for B-Raf, is highly associated with breast cancer. Glycine mutations in the glycine-rich region of B-raf are commonly found in tumor cells, providing a possible explanation for BRAF's cancer-causing potential.[13]

Bad Blood

Red blood cells make up the biggest component of blood. They are unusual in several ways compared to other cells. They ditch their nuclei early in their maturation process. Since mature red blood cells have no DNA, they're unable to synthesize new proteins. Red blood cells also have a relatively short lifespan. They're continuously renewed as new arrivals come out of the bone marrow, and they're constantly cleared in the spleen as they grow old, after about a hundred days. Their main function is to distribute oxygen to the tissues. They carry oxygen bound to iron in their hemoglobin, a transport protein they have in abundance.

Red blood cells are also unusual in that they have no mitochondria. Mitochondria are the organelles that maintain the energy supply to a cell in the form of ATP. Because red blood cells have no mitochondria, they must get their ATP from another source. For this they use glycolysis, an anaerobic process that doesn't depend on oxygen. Glycolysis produces a small amount of ATP from glucose molecules by converting glucose to lactate. Ironically, red blood cells have a lot of oxygen available to them but they don't use it to oxidize glucose. This avoidance may be a strategy to help reduce their exposure to reactive oxygen species and minimize their need for antioxidant defenses. (During normal metabolic processes, the cells in our bodies produce a type of unstable small molecule containing oxygen that reacts with other molecules in a cell. This small molecule is called a free radical, or a reactive oxygen species.) It might also not be "intelligent" to consume the resource (oxygen) that they're responsible for supplying to the tissues.

Red blood cells use an enzyme, pyruvate kinase, that is important in the glycolysis pathway. Pyruvate kinase deficiency due to genetic mutations is the second most common cause of hemolytic anemia, a condition in which red blood cells are destroyed faster than the bone marrow produces them.[14] This enzyme executes a major step during glycolysis where an ATP molecule is synthesized from ADP in order to generate energy for the cell. Generating energy via glycolysis is particularly important for red blood cells because they don't have any mitochondria.[15] Without mitochondria, red blood cells depend almost entirely on glycolysis to meet their energy needs. When the red blood cell runs out of energy, all active processes must stop. The red blood cell starts leaking potassium because its potassium pump has no energy. This loss of electrolytes causes water to leak out, as well. The cell shrinks, ultimately dying from dehydration.

Just like EPSP synthase, pyruvate kinase binds PEP. Its job is to transfer phosphate from PEP to ADP, to yield ATP. As you know, it is the PEP binding site in EPSP synthase that glyphosate disrupts. In addition to binding both PEP and ADP, which both contain phosphate, pyruvate kinase also binds to fructose diphosphate (fructose-P-P). The point is that pyruvate kinase has many phosphate-binding requirements. All of these sites could be disrupted through glyphosate substitution for glycine.

Over 50 different genetic mutations in pyruvate kinase have been linked to hemolytic anemia. Several of these mutations involve changing glycine to something else. One way to find out if a particular amino acid plays a critical role in a protein is to look at the peptide sequence for the different versions produced by different species. One such study compared the human sequence with an enzyme produced by three other species (rats, rabbits, and cats). Researchers found that there were 11 completely conserved glycine residues in the peptide sequence that were present in all four species. No other amino acid had anywhere close to this many perfect alignments.[16] This suggests, of course, that the amino acid glycine plays a critical role in pyruvate kinase. I'd further suggest that glyphosate substitution for glycine in the enzyme pyruvate kinase is particularly damaging to red blood cells.

Antioxidant Defenses

Red blood cells also metabolize glucose through another pathway, parallel to glycolysis, called the pentose phosphate pathway (PPP). The pentose phosphate pathway is critical for protection from oxidative damage. You may have heard of glutathione, an excellent antioxidant that is produced by the liver and helps the body rid itself of toxins. Think of glutathione as nature's mop. However, glutathione works as an antioxidant only in its reduced state, as GSH. That is, like other antioxidants, glutathione "works" by getting oxidized and in this way defuses reactive oxygen species. Oxidized glutathione (GSSG) has to be converted back to two molecules of reduced glutathione (2GSH) by stealing hydrogen atoms from an extremely important molecule that participates in a huge number of reactions involving oxidation and reduction. This molecule, called nicotinamide adenine dinucleotide phosphate (NADPH) is derived from vitamin B_3 (niacin) or from tryptophan. Both of these, in turn, are derived from the shikimate pathway, which glyphosate disrupts. NADPH itself has to be regenerated from NADP+ to help glutathione remain in its reduced state. NADPH contains three phosphate anions. Are you still with me?

This chain of reactions goes one step further to enable vitamin C (ascorbate) to work as an antioxidant. Reduced glutathione can react with oxidized ascorbate to convert *it* to a beneficial form that has antioxidant properties that the body can use. Oxidized ascorbate without the help of glutathione can actually be dangerous. It can be metabolized to oxalate, which increases the oxalate burden and the risk of kidney stones.[17]

Another way red blood cells can "break badly" is via the glucose-6-phosphate dehydrogenase enzyme (G6PD), which directly impacts glutathione and ascorbate. G6PD is the rate-limiting enzyme of the pentose phosphate pathway. The rate-limiting enzyme of a pathway is the slowest enzyme in the pathway, and often the one most regulated by external signaling. Red blood cells have large quantities of G6PD. They use it to renew all the antioxidant defenses. G6PD restores NADPH from NADP+. Besides its role in restoring glutathione and vitamin C to their beneficial state, NADPH is essential for the synthesis of fatty acids, cholesterol neurotransmitters, nucleotides, and amino acids. The cell expends a significant amount of energy keeping NADPH in its reduced form.

Humans can have many different genetic variants of G6PD, some of which may be defective, leading to disease. In fact, G6PD is the most heavily mutated protein in the human genome. Both G6PD deficiency and glutathione deficiency increase the sensitivity of hemoglobin to glycation damage from elevated blood sugars.[18] (Glycation is a nonenzymatic reaction that causes sugars such as glucose and fructose to attach themselves to proteins such as hemoglobin. They can also attach to fatty acids.) A common test for glucose intolerance involves testing for hemoglobin A1c (glycated hemoglobin) in the blood, as disruption of G6PD is tied to glucose intolerance.

Severely defective variants of G6PD cause neonatal jaundice, which can lead to brain damage due to high levels of a toxic byproduct of hemoglobin metabolism.[19] G6PD deficiency has also been linked to autism.[20] Because the gene is located on the X chromosome, boys are more susceptible to G6PD deficiency than girls. Boys are also nearly four times as likely to have neurological disorders, particularly autism, than girls. G6PD deficiency has also been linked to bipolar disease, schizophrenia, erectile dysfunction, and vitiligo.[21]

Glyphosate suppresses the activity of G6PD.[22] This is not surprising, since G6PD contains highly conserved glycine residues at three different sites where it binds phosphate. The enzyme binds to G6P as well as to two different molecules of NADP+.[23] A glycine residue at location 488 is within one of the NADP+ binding sites. Mutations to a bulkier amino acid have been shown to decrease the binding affinity to NADP+.[24]

We know from a study published in 2015 that the activity of the common enzyme catalase is reduced by 40 percent in people with autism.[25] Catalase is an important antioxidant enzyme in the red blood cells, the liver, and throughout the body. Catalase contains a heme group, and glyphosate has been shown to interfere with the synthesis of pyrrole, the basic unit in the porphyrin ring in heme. The first step in the synthesis of pyrrole involves glycine as a substrate, and glyphosate's role as a glycine imposter disrupts this step.[26] Catalase also binds to NADPH, and this binding is essential to keep catalase from being inactivated by its highly reactive substrate, hydrogen peroxide.[27] Glyphosate's disruption of both the supply of NADPH as well as NADPH binding could interfere with catalase's ability to protect itself from getting destroyed by the reactive oxygen species, hydrogen peroxide, that it is trying to detoxify.

Oxidative Stress and Disease

I mentioned earlier in this chapter that our cells produce reactive oxygen species, or free radicals, during normal metabolic processes. At the same time, our cells also produce antioxidants to neutralize these free radicals. In order to stay healthy, our cells need to have a balance between these free radicals and antioxidants. Oxidative stress happens when there are too many free radicals for the antioxidants to neutralize. Exposure to environmental pollution and radiation can contribute to oxidative stress.

Oxidative stress, which has been linked to neurological problems in children and a host of other disorders, is associated with a decreased ratio of reduced glutathione versus oxidized glutathione, impaired catalase activity, mitochondrial dysfunction, damage to cell membrane fatty acids, and neuroexcitotoxicity.[28] It's imperative that we research whether glyphosate disruption of multiple proteins that bind NADPH is leading to oxidative stress. Since studies on human skin cells have shown that a glyphosate-based formulation disrupts antioxidant defenses in multiple ways and induces oxidative damage to membrane fatty acids, it seems likely that glyphosate exposure causes oxidative stress.[29] Glyphosate exposure decreases glutathione levels in the blood and in the brain, heart, liver, and kidneys of rats, while significantly increasing a common marker for oxidative damage.[30]

Glyphosate disrupts antioxidant defenses in plants as well as in animals. Genetically modified glyphosate-resistant soybeans exposed to glyphosate experience a sharply decreased ratio of reduced glutathione to oxidized glutathione in their leaves.[31] In both plants and animals, maintaining glutathione in its healthy reduced form depends on having an adequate supply of NADPH, which in turn depends on a functioning G6PD enzyme. These studies suggest that glyphosate disrupts enzymes that maintain NADPH in its reduced form (as NADPH instead of NADP+). It is plausible that this effect on plants is also due to glyphosate disrupting NADPH binding in their antioxidant defense enzymes.

Methylation Pathways

The enzyme methylenetetrahydrofolate reductase (MTHFR) plays a critical role in methylation pathways. Methylation is the attachment of a methyl group, CH_3, to a molecule. MTHFR binds to NADPH and has a highly conserved GxGxxG motif beginning at residue 498, instantiated as GWGPSG. Mutations at four other glycine residues also cause a reduction in enzyme activity.[32] Two of these are substitutions by aspartate, one of the two negatively charged coding amino acids, a good model for glyphosate substitution. Methylation pathways play a critical role in development where the methylation pattern in the genes dictates metabolic policy. Methylation is one of the key contributors to epigenetics, something I'll talk more about in chapter 8, and patterns can be passed down across multiple generations.

Methylation pathway impairment is commonly observed in children with autism.[33] Dr. Mostafa Waly, a biochemist and specialist in prevention of chronic diseases in the Food Science and Nutrition Department at the Sultan Qaboos University in Oman, has proposed that a central feature of autism is altered epigenetics due to disrupted methylation pathways. These pathways are modified in response to oxidative stress due to impaired antioxidant defenses.[34] This observation is consistent with what we have seen in our model of the glyphosate susceptibility motif, which would impact enzymes that are critical for both methylation pathways and antioxidant defenses, due in part to disruption of their ability to bind NADPH.

Impaired Liver Function

The liver is responsible for breaking down and clearing many compounds, including steroid hormones, from the body. Liver function is dependent upon a group of enzymes called cytochrome P450 (CYP) enzymes. CYP enzymes play a crucial role in metabolizing drugs, as well as in synthesizing cholesterol, steroid hormones, fatty acids, and eicosanoids, which are signaling molecules derived from fatty acids.[35] Mutations in CYP enzymes are linked to disordered steroidogenesis, ambiguous genitalia, and Antley-Bixler syndrome, a rare condition characterized by deformities affecting the skeleton.[36]

CYP reductase is an enzyme that converts NADP+ in CYP enzymes to NADPH, a necessary step before they can do their job. Multiple species, from microbes to plants to insects to mammals, have this enzyme.[37] All CYP enzymes are impaired when CYP reductase is impaired. Because of its crucial role, complete knockout of the CYP reductase gene is lethal to mouse embryos. When mice are genetically engineered to have a minimally functioning variant of CYP reductase in their livers, they suffer severe impairment. They become unable to clear drugs from the body and they build up fat and cholesterol in their livers.[38] Rats chronically exposed to very low doses of glyphosate also develop nonalcoholic fatty liver disease, something we'll be talking more about in chapter 7.[39]

Flavin mononucleotide (FMN) and flavin adenine dinucleotide (FAD) are members of a class of molecules called flavins. FMN contains one phosphate and FAD contains two. There is a large class of enzymes called flavoproteins that are defined by the fact that they bind to flavins. CYP reductase is an important member of this class, and its binding to FMN is essential for it to work. In 44 different species that possess CYP reductase, glycine residues are conserved at the FMN binding site.[40] At least one of these highly conserved glycine residues binds with phosphate, consistent with the glyphosate susceptibility motif:[41] (1) The protein binds phosphate at a site where at least one glycine residue is highly conserved; (2) At least one glycine residue at this site has a small amino acid immediately to the left; (3) There is at least one, and preferably two or more, positively charged amino acids nearby that can securely hold glyphosate's methylphosphonate group in place. We know that glyphosate suppresses activity of liver CYP enzymes in rats.[42] One likely explanation for this is that CYP reductase in the liver is disrupted by substitution of a critical glycine at the phosphate binding site for FMN.

Malfunctioning White Blood Cells

NADPH oxidase is an important enzyme in the body. It has many roles, one of which is to kill pathogens. Chronic granulomatous disease is a rare genetic disease where immune cells have a defective variant of NADPH oxidase.[43] As a result, the immune cells are unable to synthesize superoxide, a powerful oxidizing agent that keeps pathogens, such as the fungus *Aspergillus fumigatus*, as well as *Staphylococcus aureus*, *Listeria*, *Klebsiella*, and *Pseudomonas*, in check.

Those living with chronic granulomatous disease can develop bacterial and fungal infections in their brains, intestines, livers, lungs, lymph nodes, spleens, stomachs, and on their skin. Lung infections, especially fungal pneumonia, are common. So are granulomas, areas of inflamed tissue, which is where the disease gets its name. People born with this genetic defect often die before age 40 from uncontrollable infections.

Chronic granulomatous disease usually involves mutations in glycine residues at the phosphate binding site of the enzyme NADPH oxidase. NADPH oxidase binds NADPH at a site that has three glycine residues. Swapping out the first glycine of the GxGxG sequence for a bulkier, negatively charged molecule such as glutamate (or glyphosate) not only prevents NADPH from entering its active site but also disrupts FAD binding, abolishing enzymatic activity.[44] Glyphosate exposure mimics this genetic disease by disrupting NADPH oxidase activity in the same way the genetic mutation that causes chronic granulomatous disease does.

In chapter 2, I discussed *Candida auris*, a deadly new fungal infection that is practically untreatable, and how glyphosate's disruption of our immune defenses is contributing to the rise in life-threatening fungal and other pathogenic infections. A genetic mutation in a glycine residue, especially if it is substituted by aspartate or glutamate—two amino acids that are similar to glyphosate in that they are bulkier than glycine and have a negative charge—models the kind of scenarios we can expect to see when glyphosate substitutes for glycine. In other words, the phenomena of chronic granulomatous disease helps explain how glyphosate impairs immune function. Several other proteins in addition to NADPH oxidase are involved in immunity. It can be predicted that these other proteins would be disrupted by glyphosate, something I'll be talking more about in chapter 10.

Sulfur Sensitivity

Some people claim they can't eat sulfur-containing foods because they have a sulfur sensitivity. Often, this is because they have an overgrowth in their gut of bacterial strains that reduce sulfite and sulfate to hydrogen sulfide (H_2S) gas. This sensitivity can cause bloating, gas pain, and brain fog because hydrogen sulfide gas freely floats through tissues and can reach the brain and disrupt its function. Involved are two distinct kinds of an enzyme called sulfite reductase that reduces sulfite. One enzyme incorporates sulfur into an organic molecule. The other enzyme is content to just make hydrogen sulfide gas.

The former, assimilatory sulfite reductase, produces cysteine and methionine as organic products. It is a "hemoflavoprotein," meaning that it depends on both NADPH and FAD (a flavin) binding, and also attaches to heme. Since NADPH and FAD are phosphorylated molecules, assimilatory sulfite reductase is likely to be disturbed by glyphosate.[45] As we've seen, synthesis of heme is also disrupted by glyphosate. By contrast, dissimilatory sulfite reductase, which produces hydrogen sulfide gas, has none of these dependencies. Might glyphosate's disruption of the former in the gut microbes be what leads to overproduction of the latter? If this were the case, it would also allow the sulfur-reducing bacteria associated with autism, and that play a causal role in allowing inflammatory bowel disease (IBD), to thrive.[46]

Detecting Sequences

In chapter 4, I referred to a team of scientists who claimed to prove that glyphosate does not substitute for glycine in protein synthesis.[47] The team, led by Michael Antoniou, PhD, a molecular geneticist at King's College London, designed their experiment explicitly to see whether glyphosate is substituting for glycine during protein synthesis, and they evaluated this on breast cancer cells grown in an in vitro culture. The team concluded that all of the "hits" they found—places where the technology detected a glyphosate substitution for glycine in a specific protein—were false positives, because their "unexposed" cells (the controls) also showed evidence of glyphosate substitution. As we talked about earlier, what Antoniou's group failed to realize was that the "unexposed" cells had likely been exposed to glyphosate over an extended time period in advance of the experiment, initially in the human body, and subsequently during many generations of growth in vitro, including during the experiment itself, when they were supplied with glyphosate-contaminated nutrients.

Fortunately, their paper provided the exact sequences of the 15 proteins they detected, and I was able to conduct analysis using the Universal Protein Resource (UniProt) website's biological sequencing software. Through UniProt, I was able to identify all 15 proteins, at least 9 of which bind to phosphate-containing molecules (see table A.3 on page 187). This finding strongly supports my hypothesis that proteins with glycine residues that bind to phosphate are especially susceptible to glyphosate substitution.

Identifying 15 human proteins that appear to have been modified through glyphosate substitution for a specific glycine residue may be a significant breakthrough in the quest for a test to detect glyphosate contamination in proteins. Although I refute the team's conclusions in the strongest possible terms, I also acknowledge that their research is of great value because it specifies a prescribed procedure that can be applied to other cell types grown in culture, as well as to biological samples extracted from diseased tissues in mammals, such as fingernails of scleroderma patients, skin cells of psoriasis patients, amyloid beta plaque from Alzheimer's patients, hair samples of children with autism, hooves of horses suffering from founder, biopsies from cancer tumors, and diseased kidney and liver tissues.

As we collect a database of specific substitution patterns, we may even be able to predict rules for peptide contexts where glycine residues are especially susceptible, such as when neighboring amino acids are small or positively charged. These patterns are already becoming apparent in the small set retrieved in Antoniou's experiment. Six of the 15 purported substituted glycines were immediately followed by a positively charged amino acid: lysine, histidine, or arginine, and 10 were immediately preceded by one of valine, leucine, serine, or threonine, all of which are small amino acids, supporting space for glyphosate's methylphosphonyl tail.

— CHAPTER 6 —

Sulfate: Miracle Worker

This ability to essentially "breathe" sulfur compounds has long been thought to be one of the earliest stages in the transition from a non-biological to biological world.

—David Wacey, PhD

You may never have given much thought to sulfate, but it plays many important roles in maintaining your health. A molecule present in all living tissues, sulfate does many good things in your body; here are some highlights. It escorts hormones through the bloodstream from one place to another, essentially turning the hormones off until they arrive where they need to go. It deactivates toxic elements and compounds, including mercury and acetaminophen, and shuttles them out of the body. Sulfate helps maintain the acidity of cell organelles, such as the lysosomes, the cell's digestive system. Sulfate is also essential for maintaining a healthy barrier in blood vessels to keep the vessel wall shielded from reactive substances in the blood and to keep the water in the blood from leaking out into the tissues. As cholesterol sulfate in red blood cell membranes, it creates a negative charge that causes red blood cells to repel one another, so that they don't stick together and create a no-flow situation. This negative charge also helps propel red blood cells through the capillaries. In the polysaccharide heparan sulfate, it also plays an important role in controlling what goes into and what stays out of most cells in the body.

To understand how sulfate came to be so important, we need to go on a biochemical journey together. Then we can tackle the inconvenient truth that glyphosate systemically impairs the body's ability to maintain adequate sulfate supplies, a phenomenon that is contributing to many diseases on the rise.

Essential for Life

Sulfur is an abundant and naturally occurring element. It's in brimstone, and in the smell of rotten eggs. Sulfur sits just below oxygen on the periodic table. (Elements in the same column of the periodic table have similar chemical properties, because their structures are similar.) Sulfur has many of the same properties as oxygen—and, as we all know, oxygen is essential for life. Sulfate is a combination of sulfur and oxygen, SO_4^{-2}: one sulfur atom, four oxygen atoms, and a negative two charge. This charge allows sulfate to create a negatively charged surface when it attaches to other molecules.

When life began on our planet there was very little oxygen in the atmosphere. On early Earth, hydrogen sulfide gas from deep-sea vents and volcanic eruptions was abundant. Adenosine triphosphate, the energy currency of life, was first made with sulfur rather than oxygen.[1] Mitochondria can still oxidize hydrogen sulfide gas to make ATP using some of the same enzymes that metabolize glucose. As our cells' powerhouses, the mitochondria will preferentially take up hydrogen sulfide gas in place of glucose if it's present in the environment. Hemoglobin, which transports oxygen through the body in red blood cells, seems to have been originally designed to bind sulfur rather than oxygen.[2]

Even though we don't breathe it in as we breathe in oxygen, sulfur is crucial throughout the body. Hydrogen sulfide gas is a signaling gas synthesized in the body, telling cells to initiate certain processes. Other signaling gases that you might be more familiar with include nitric oxide and carbon monoxide. In small amounts hydrogen sulfide is beneficial. But, like carbon monoxide, it can also kill you in higher amounts.

Not Just a Shuttle Service

Sulfation is the process of attaching a sulfate anion to another molecule. This process plays multiple important roles in biology. For example, sulfate attaches to fat-soluble toxins and toxic chemicals in the liver to make them water soluble so that they can be excreted through the urine. Sulfate also attaches to all the hormones that are produced in the adrenal glands before they are shipped out into circulation. Melatonin, the hormone that helps you sleep, is sulfated after it is produced by the pineal gland. Cholesterol is sulfated in the skin to create a healthy skin barrier. Specialized enzymes called sulfotransferases move sulfate from a source molecule—usually a modified form of ATP, called phosphoadenosine phosphosulfate (PAPS)—to another molecule. PAPS is the energy form of sulfate, similar to ATP except with a sulfate instead of one of the phosphates.

PAPS synthase is the enzyme that converts sulfate into PAPS by binding to two different ATP molecules. One is used as a source of energy *for* the reaction. The other one will be turned into PAPS *by* the reaction. PAPS synthase has two highly conserved amino acid motifs associated with these two ATP binding sites, and each of these motifs contains three glycines. PAPS synthase can be disrupted by glyphosate substituting for glycine at either of these two sites.

Sulfate, Steroids, and Sunlight

Steroid hormones are an important class of hormones that the body uses to maintain health. They include the stress hormone cortisol produced by the adrenal glands and the sex hormones produced by the adrenal glands, testes, ovaries, and placenta. Attaching a sulfate temporarily inactivates the hormone during transit and makes it more water soluble so it can easily travel through the bloodstream. In other words, sulfate is essential for distributing hormones throughout the body.

However, sulfation also inactivates the *sulfate* as it travels through the body. This is crucial for distributing sulfate to the tissues. Too much free sulfate (not attached to a carrier molecule) would change the state of the water in the blood, making it too thick and impeding blood circulation through the bloodstream. Biological organisms have come up with an elegant solution: hooking sulfate onto carrier molecules, such as hormones.

For example, the adrenal glands, which are located on top of each kidney, produce dehydroepiandrosterone (DHEA) sulfate. If you're a woman in menopause, you may have been prescribed DHEA by your doctor. That's because women entering menopause can become low in testosterone, and DHEA can be converted into testosterone, lessening some symptoms associated with menopause.[3] DHEA sulfate also plays an important role during pregnancy. It is converted by the placenta into estrogen to help orchestrate fetal development. The sulfate is crucial; the unsulfated form of DHEA cannot substitute for the sulfated form. This is particularly true in the later part of pregnancy when the baby's brain is maturing. Insufficient supplies of DHEA sulfate can result in anencephaly, a baby born without a cerebral cortex.[4]

Leydig cells in the testes are responsible for supplying testosterone to sperm. An in vitro study on a tumor cell line derived from Leydig cells determined that exposure to Roundup decreased their production of testosterone by 94 percent! The enzymatic action of the cytochrome P450 (CYP) enzymes in these cells was suppressed, as well as the activity of a protein that is responsible for transporting cholesterol across the mitochondrial membrane to deliver it to the CYP enzymes.[5] CYP enzymes synthesize steroids from cholesterol. Their disruption by glyphosate could disrupt the entire endocrine system.

Cells in the skin produce cholesterol sulfate in response to sunlight, and this, too, circulates in the blood at relatively high concentrations. Vitamin D, which is actually a hormone, is derived from cholesterol, triggered by sunlight, and commonly sulfated in transit. Children with autism and other neurological disorders often have low serum cholesterol,[6] low serum sulfate,[7] and low serum vitamin D.[8] This suggests that cholesterol, vitamin D, and sulfate are all important for brain health, and that sunlight exposure can help protect the brain. (It also suggests that slathering children in sunscreen may have unintended consequences that we are only beginning to appreciate.)

Sulfate and Neurotransmitters

As you know from the introduction, plants and gut microbes produce the aromatic amino acids tryptophan, tyrosine, and phenylalanine in the shikimate pathway, which glyphosate disrupts. Tryptophan is a precursor to the neurotransmitters serotonin and melatonin. Phenylalanine and tyrosine are precursors to dopamine, adrenaline, and the skin's tanning pigment melanin. Thyroid hormones that regulate metabolism are also derived from tyrosine. All of these hormones, as well as tryptophan itself, are commonly sulfated in transit. And all of these molecules are important biologically. Their disruption can have dire effects.

Many nutrients that offer protection from conditions like Alzheimer's disease, heart disease, arthritis, and cancer are sulfated in transit. These include resveratrol found in grapes and wine; curcumin, an anti-inflammatory compound found in turmeric; vitamin C (ascorbic acid); and various beneficial polyphenols and flavonoids found in colorful fruits, vegetables, coffee, tea, and chocolate. I believe the benefits of these substances are largely derived from their ability to transport and deliver sulfate to the tissues. All of these nutrients are synthesized by plants and derived from products of the shikimate pathway.

Glyphosate's disruption of steroid hormones along with its suppression of the synthesis of aromatic amino acids and their derivatives (neurotransmitters and polyphenols) can interfere with the body's ability to transport and deliver the sulfate that is normally carried by these bioactive molecules.

Water Is Life

The Tuareg people of the Sahara have a proverb, *"aman iman,"* meaning, "water is life." Everyone knows we can't live without water, but few people realize just how special water molecules are biophysically. Water seems simple but it's not. Water science is dense and complex. One of its most important properties is the one that is least well understood. Beyond solid, liquid, and gas, water has what Gerald Pollack, PhD, a scientist in the Department of Bioengineering at the University of Washington, calls a "fourth phase."

Also called gelled, structured, ordered, liquid crystalline, exclusion zone, or living water, the fourth phase of water is highly organized and electrically conductive, but not quite a solid. Think of the water in gelatin desserts. Water molecules are very small. The human body is over 98 percent water by molecule count but only 66 percent by mass. Most of the water in our bodies is in the fourth phase. Except for the water in our blood. The water in our blood needs to be fluid in order for our blood to flow smoothly through the blood vessels.

Sulfate is essential for maintaining the fourth phase of water. When a water-soluble surface is populated with attached sulfate anions, the water adjacent to the surface actually organizes into a regular array of hexamer crystals. These six-branched crystals resemble the crystalline structure of ice, creating a firm gel instead of liquid water.

For example, blood vessels are encased in a single layer of cells called endothelial cells that make up the endothelium, or vessel wall. The endothelial cells are layered onto proteins embedded in the plasma membrane of each cell along the length of the endothelium. Polysaccharides, or chains of sugar molecules, attach to the proteins. Sulfate anions then attach to these polysaccharides in an irregular pattern. The resulting large and complex molecules are called sulfated glycosaminoglycans, consisting of sulfate, glucose (glycos-), nitrogen (amino-), and polysaccharides (glycans).

Chondroitin sulfate is a sulfated glycosaminoglycan found in joint cartilage. People who suffer from arthritis and joint pain find that taking chondroitin in supplement form can help. Heparan sulfate is another common sulfated glycosaminoglycan. Constantly synthesized and broken down, heparan sulfate lasts only a few hours, during which time it temporarily binds to proteins embedded in cell membranes, "snags" contents from the blood that will be useful to the cell (such as lipoproteins), and facilitates almost all the cell's core activities.[9] Many signaling molecules, particularly growth factors, bind to heparan sulfate to initiate their signaling response.

When heparan sulfate detaches from the proteins in the cell membrane, it travels into the cell—along with whatever is attached to it—to the lysosome, the cell's digestive system. The sulfate anions then detach from heparan sulfate in the lysosome, and become sulfuric acid, which helps maintain the low pH essential for the lysosome to digest nutrients and break down cellular debris. When sulfate becomes deficient, a cell's ability to clear cellular debris for recycling is impaired, leading to an accumulation of garbage that can't be removed.

Heparan sulfate is also the most abundant glycosaminoglycan—chondroitin sulfate is another—that forms the glycocalyx, a structured fourth phase "extracellular matrix," that surrounds and protects healthy cells. The glycocalyx also forms a structured fourth phase "wall" that lines all blood vessels in the body, forming an almost impermeable barrier to reactive substances carried in the blood. The viscosity of water bound to hydrophilic surfaces, such as the sulfated glycocalyx, is orders of magnitude higher than the viscosity of the water in the flowing blood.[10] It's crucial for the glycocalyx to have sufficient sulfate in order to maintain a healthy barrier.[11] In other words, if glycosaminoglycans are insufficiently sulfated we're in trouble.

Dr. Pollack and his team conducted experiments on the effects of biologically active substances on this exclusion zone using a rough simulation of a glycocalyx. They determined that certain foods and spices, including coconut water, turmeric, and basil, increased the thickness of the barrier up to a certain concentration, suggesting that these foods may be good for protecting the integrity of the blood vessels. On the other hand, the glyphosate formulation Roundup, even at very low concentrations, decreased the thickness of the barrier.[12] A thinner barrier can be more easily breached by toxic or highly reactive substances in the blood.

A remarkable property of structured water is that it extrudes protons. The gel becomes negatively charged, creating a sort of battery that can supply electricity to the body. Like the surface of Jell-O, the gelled water provides a smooth, slick surface next to the liquid water in the blood. A slick surface is especially important in the capillaries to make it almost effortless for the red blood cells to slide through with minimal friction. Insufficient sulfate in the capillary walls increases the resistance to blood flow, causing extra load on the heart. When someone has a long-term sulfate depletion, we see these kinds of vascular and cardiac problems.

Sulfate and Phosphate Similarities

Sulfate shares biological similarity with phosphate. Both molecules consist of a central atom surrounded by four oxygen atoms. In phosphate, the central atom is phosphorus. In sulfate, the central atom is sulfur. Both sulfate and phosphate have a negative two charge.

As we've seen, heparan sulfate consists of long chains of polysaccharides. Heparin is a biologically important molecule that is similar to heparan sulfate. Heparin is released from mast cells under inflammatory conditions, such as allergic reactions. Although heparin does not contain the word *sulfate* in its name, it is actually more highly sulfated than heparan sulfate. In fact, heparin is the most highly sulfated molecule in biology. There are many important proteins that bind to heparin or to heparan sulfate. And as with phosphate, it can be surmised that some of these will be impacted by glyphosate substitution.

Unlike phosphate-binding sites, heparan-sulfate-binding sites cannot be characterized by a simple amino acid motif. In fact, researchers are still trying to identify the specific site in different proteins where heparan sulfate binding takes place. One study designed to characterize features of heparin-binding sites and heparan-sulfate-binding sites determined that glycine, as well as the positively charged amino acids lysine and arginine, were overrepresented at sites that bind to heparin or heparan sulfate.[13] So, despite the fact that the pattern cannot be characterized by a simple motif, the same principle that applies to phosphate-binding sites may also apply to heparin- and heparan-sulfate-binding sites, making it likely that these sites will be sensitive to glyphosate substitution in the same way.

Disrupted Signaling Mechanisms

Many cytokines, including interferons, interleukins, and growth factors, bind to heparan sulfate. Such binding retains them close to the place where they are released from a cell, restricting their activity to what's called a paracrine (local) effect rather than an endocrine (global) effect. That is, the binding keeps them from getting out into the body's general circulation. If the binding to heparan sulfate is weak, the cytokine effect will no longer be localized, and the cytokines will behave erratically.

Several biologically active proteins bind to heparan sulfate in the glycocalyx (see table A.4 on page 188).[14] This binding is necessary to transduce their signal to the cell so that it can react appropriately. These proteins play important roles in controlling blood clots, stimulating cell division, maintaining cell adhesion to a surface, and stimulating growth, as well as inducing an appropriate inflammatory response.

The proteins listed in the appendix all have critically important roles in the body. The fibroblast growth factors (FGFs) are a family of proteins that play an essential role in development. FGF binding to heparan sulfate is essential for engagement with the FGF receptor in the membrane. Tissue transglutaminase binding to heparan sulfate prevents it from cutting loose and causing fibrosis (scar tissue) by building many crosslinks in collagen. Interferon binding to circulating heparin prevents it from being degraded by proteolysis enzymes.[15] Another excellent example of a protein that binds heparin and heparan sulfate is antithrombin. This protein protects the body against blood clots by inactivating several enzymes involved in the blood clotting cascade. In fact, heparin is commonly administered to protect from blood clots during surgery. Heparin binding by antithrombin increases antithrombin's activity by a thousandfold.[16]

If proteins are impaired in their ability to bind heparin or heparan sulfate, various defects may occur in the basic biological mechanisms essential for homeostasis. Proteins that can't bind heparin due to glyphosate incorporation will end up wandering free in the blood and getting cleared too rapidly by digestive proteins. Proteins that can't bind heparan sulfate in the cellular membranes will also wander into the bloodstream, causing an unintended and unpredictable systemic response. It's difficult in the abstract to predict the consequences of such basic disruption of biological mechanisms, but we can hypothesize that it can lead to signaling responses causing systemic inflammation, hemorrhaging, thrombosis, and more.

Smooth Flow

Endothelial nitric oxide synthase, or eNOS, is an important enzyme that synthesizes nitric oxide, a signaling gas that relaxes the vessel wall, diffuses into the blood, and promotes blood flow. Impaired nitric oxide synthesis is associated with hypertension, cardiovascular disease, and diabetes.[17] eNOS produces nitric oxide only when it is inside the cell, phosphorylated, and bound to a protein called calmodulin. In its "inactive" mode, eNOS stays tightly bound to a protein called caveolin in the cell's membrane.

I believe that synthesizing nitric oxide is only one of eNOS's important roles. Its other role is to synthesize sulfur dioxide (SO_2), which can be further metabolized to sulfate by the enzyme sulfite oxidase. That is, if I am correct, eNOS is not only essential for synthesizing nitric oxide. It is also essential for maintaining adequate supplies of sulfate. I also suspect, although this is admittedly more speculative, that eNOS synthesizes sulfur dioxide when it is bound to the membrane—not inside the cells, as with nitric oxide synthesis—and that eNOS relies on sunlight to catalyze the reaction.[18]

When calcium enters an endothelial cell, it can launch a signaling cascade that causes eNOS to detach from the membrane, enter the cell, phosphorylate, bind to calmodulin, and switch to synthesizing nitric oxide. In the presence of oxygen, nitric oxide oxidizes to nitrite, which can erode the glycoproteins that form the glycocalyx. (This is why the heparan sulfate survives for only a few hours in the glycocalyx: The nitric oxide produced by eNOS facilitates the recycling of the glycocalyx heparan sulfate.) In other words, eNOS alternates between the synthesis of sulfur dioxide, which is oxidized to sulfite and sulfate, and the synthesis of nitric oxide, which is oxidized to nitrite and nitrate. Sulfate builds up the extracellular matrix. Nitrite dismantles it. I think of it as a delicate yin and yang balance.

A Mystery in Our Red Blood Cells

Red blood cells contain a lot of eNOS, and this fact poses a big mystery. Red blood cells are good at keeping calcium out, suppressing the signal that would prompt eNOS to detach from the membrane, enter the cell, and synthesize nitric oxide. In fact, the eNOS in red blood cells never *leaves* the cell membrane.[19] Arginine is needed for eNOS to make nitric oxide. Not only do red blood cells not possess the transport protein that imports arginine into the cell, but they also contain arginase, an enzyme that degrades arginine.

So, despite having an abundance of the enzyme that synthesizes nitric oxide (eNOS), red blood cells have virtually none of the other requirements. It appears as if red blood cells are doing everything in their power to make sure eNOS never accomplishes anything. It would not be a good idea for red blood cells to synthesize nitric oxide anyway. The nitric oxide would bind to hemoglobin and interfere with the ability of red blood cells to transport oxygen, much like carbon monoxide does. So the big question is: *Why* do red blood cells produce so much eNOS?

Here's what I think: We know that the endogenous steroid cholesterol sulfate is important to red blood cells. It helps provide the negative charge in the cell membrane that keeps cells from glomming together and blocking blood flow. We also know that the enzyme eNOS oxidizes nitrogen to nitric oxide when it is inside the cell. It is plausible that red blood cells carry eNOS because eNOS can oxidize sulfur to sulfur dioxide when it is attached to the membrane, providing the raw materials for sulfating the cholesterol.

eNOS is part of a family of enzymes knowns as flavoproteins, which play an important role in many biological processes. As we've seen, flavoproteins are susceptible to glyphosate substitution at the site that binds phosphate-containing molecules. eNOS binds both NADPH and FAD. eNOS also binds heme, the iron-binding oxygen-carrying molecule in hemoglobin. As I explained in chapter 5, heme synthesis is impaired by glyphosate. That's not all. eNOS depends on both iron and zinc as catalysts. In the presence of glyphosate, iron and zinc may become deficient due to glyphosate's chelation. Finally, eNOS contains a terminal glycine residue that is essential for its binding to the cell membrane in a process called myristoylation.[20] A terminal glycine is an easy place to substitute glyphosate, since there is plenty of room for its methylphosphonate unit—with no amino acid to the right. These features show many vulnerabilities to glyphosate through metal chelation, disruption of heme synthesis, phosphate binding at the flavin site, and disruption of myristoylation and membrane-binding.

eNOS has a reputation among scientists for its capacity to misbehave. Literally thousands of articles have been published on eNOS dysfunction. For example, eNOS will sometimes spew out reactive superoxide instead of producing nitric oxide.[21] I propose that glyphosate disrupts eNOS's functionality, and this not only causes oxidative damage to the plasma membrane but also interferes with sulfate supplies to the body. The superoxide that should have been consumed by catalyzing the synthesis of sulfur dioxide, and, subsequently, sulfate, instead sticks around as a reactive oxygen species, because eNOS has been disabled by glyphosate.

So, briefly, this is how it goes. Red blood cells take up hydrogen sulfide gas from the bloodstream. The cell's hemoglobin converts the gas to thiosulfate.[22] The eNOS molecules in the cell membrane of red blood cells and endothelial cells of the vessel walls, together with sulfite oxidase, finish the job, transforming thiosulfate into two sulfate molecules.[23] Flavins bound to eNOS act as a photo-initiator, responding to blue light by releasing electrons that can then catalyze the reaction to oxidize sulfur. In other words, eNOS synthesizes sulfate in response to sunlight. I would even surmise that this is a far more important benefit of sunlight exposure than the synthesis of vitamin D. This could also explain why red blood cells have a lot of eNOS in their membrane: They use it to oxidize sulfur, not nitrogen.

High serum cholesterol is a well-known risk factor for cardiovascular disease. Several experiments have shown that high serum cholesterol, also known as elevated LDL, induces endothelial cells to produce more of the protein caveolin, which keeps eNOS at the membrane. This changes the behavior of eNOS, causing it to release superoxide, in a pathological situation referred to as eNOS uncoupling.[24] This is very dangerous. Superoxide reacts with nitric oxide to produce peroxynitrite, one of the most reactive substances known to biology. I believe this pathology is a consequence of glyphosate's disruption of eNOS's ability to oxidize sulfur to produce sulfate. Poor binding to the flavin due to glyphosate's obstruction prevents eNOS from completing the reaction that combines sulfur with superoxide to produce sulfur dioxide. Instead, the superoxide gets released and wreaks havoc.

The cholesterol, meanwhile, is trying to signal endothelial cells to produce sulfate that can pair with cholesterol to produce the water-soluble molecule cholesterol sulfate. Cholesterol sulfate can then enter the membranes of the LDL particles to protect them from oxidation and glycation damage. But none of this works well when glyphosate is disrupting eNOS. LDL gets oxidized. Oxidized LDL that accumulates in the arteries is the harbinger of cardiovascular disease.[25]

Finding Balance

High levels of free sulfate in our blood would cause it to gel and reduce the flow of blood. So the body must maintain low levels of sulfate in the blood within a very tight range.[26] I surmise that this is why there are so many molecules that specialize in sulfate transport. Most of the biologically active molecules that transport sulfate mentioned previously—the polyphenols and flavonoids, the aromatic amino acids and derived neurotransmitters and thyroid hormone, and the sterols like cholesterol, cortisol, vitamin D, and the sex hormones—are characterized by rings in their structure. They usually have at least one six-sided ring and one five-sided ring. I speculate that these rings disperse the negative charge on the sulfate anion and probably disrupt its ability to gel water. These rings may also facilitate the transfer of sulfate to the extracellular matrix in much the same way that flavins facilitate catalytic reactions. All these sulfate carrier molecules are likely essential for distributing sulfate via the vascular system to all the tissues, in a way that does not stop blood from flowing.

A Backup for Sulfate

Taurine is an amino acid, but a very unusual one. It has a sulfur-based (sulfonyl) group in place of the usual carbon-based (carbonyl) group. This is characteristic of the coding amino acids. But taurine is not a coding amino acid, so it never appears in proteins. It is, however, the most highly available amino acid in isolation. It's stored in particularly high concentrations in the heart, the liver, and the brain. This fact has posed a mystery to science. I believe that taurine's most important role is as a stored form of sulfate, readily available in an emergency should sulfate levels become dangerously low.

When sulfated glycosaminoglycans become sparse in the blood vessels, the glycocalyx becomes precariously thin, leading to a high risk of hemorrhage, or life-threatening blood clots, formed in an attempt to block potential bleeds. Curiously, the brain releases its stored taurine in large amounts during seizures. The heart does the same during a heart attack, which I view as a form of "heart seizures." These taurine molecules are picked up by the liver and conjugated to bile acids, then shipped to the gut via the gall bladder.

Taurine is considered inert, meaning that human cells cannot break it down. However, gut microbes *can* metabolize taurine to produce hydrogen sulfide gas, which can then be oxidized to make sulfate.[27] In fact, radiotracer experiments in animal models have shown that radio-labeled dietary taurine shows up later as sulfate in the urine.[28] Sulfate, produced by the gut microbes from taurine, is then available to the host to help replenish the impoverished supply in the bloodstream.

It's a little-known fact that glyphosate has been observed, experimentally, to suppress taurine uptake in *E. coli*. Taurine transporters are among the most severely affected.[29] The disruption of taurine transporter activity suggests that taurine cannot be converted to sulfate in the presence of glyphosate, resulting in widespread sulfate deficiency.

A Giant Puzzle

I love biology. Sometimes it seems like a giant jigsaw puzzle. But jigsaw puzzles are static. The puzzle pieces stay where you put them. We need to remember that biological organisms are systems and that nothing in them works in isolation. Biology, then, is more like a three-dimensional interlocking gear puzzle. We can't change anything without far-reaching effects. Oxygen is exchanged in the lungs, but if that oxygen doesn't move throughout the body, nothing works as it should. Sulfate is similar. Sulfate is arguably as important as oxygen.

I've shown you how important sulfate is to our cells' extracellular matrix, especially the endothelial glycocalyx, which lines the blood vessels. Sulfate's ability to induce gelling of water also protects cells from attack by glycating agents (also known as sugars) and oxidizing agents (reactive oxygen species that are generated by defective enzymatic reactions). It also presents a smooth surface for the red blood cells moving in capillaries, such that they can slide effortlessly with little resistance. Sulfate, in the form of cholesterol sulfate, protects lipid particles such as LDL and HDL (the so-called good cholesterol) and red blood cells from damage, keeping them healthy during circulation. It also provides negative charge that keeps the red blood cells well separated from each other, maintaining healthy blood flow.

My research has led me to conclude that glyphosate erodes the body's ability to maintain adequate sulfate, mainly through its multiple effects on proteins susceptible to glycine substitution. Glyphosate not only depletes the supply of sulfate carriers (the aromatics and the steroids), but also interferes with the synthesis of PAPS, which is essential for sulfate transfer from one molecule to another. The carriers are present in insufficient numbers, and they are unable to easily obtain a sulfate attachment. Perhaps the most devastating effect of glyphosate is the disruption of eNOS's ability to make sulfur dioxide, which normally gets oxidized to sulfate that is immediately available to populate the glycocalyx. Instead, eNOS releases superoxide, which does damage to the cellular contents. Systemic sulfate deficiency is a key factor in multiple modern diseases, including heart disease, neurological diseases, gut disorders, and autoimmunity.

— CHAPTER 7 —

Liver Disease

Of all the organs in the body the liver is the most extraordinary. In its versatility and in the indispensable nature of its functions it has no equal.

—Rachel Carson, *Silent Spring*[1]

Your heart, yes. Your brain, of course. But if you're healthy, chances are you don't spend a lot of time thinking about your liver. The largest internal organ in the body, your liver plays a crucial role in maintaining your health. Underappreciated and often misunderstood, the liver is the body's filtration system. Much of what you eat and drink eventually passes through your liver, where it is sorted into nutrients to circulate and toxicants to dispose of. It's a meaty, fleshy, reddish, rubbery organ that sits on the right side of your belly, just underneath the protective cage of your ribs. Your liver may perform as many as five hundred functions, mostly related to metabolism, vitamin and hormone regulation, and detoxification. It's the place where bile acids are produced, before storage in the gall bladder and release into the upper intestine to aid in the digestion of fats. The liver is also responsible for making sure your blood sugar doesn't get too low. It responds rapidly by releasing sugar that is synthesized from nutrients such as lactate, proteins, and fats through a process called gluconeogenesis. (If you break down the word gluconeogenesis, you can think of it as literally "making new glucose.")

The liver can also make toxic chemicals less toxic. How? By modifying them using cytochrome P450 enzymes and then conjugating the modified toxicants with small molecules like glutathione or sulfate. This solubilizes them so they can be shipped via the bloodstream to the kidneys for export in the urine. These toxic chemicals include phenolic compounds produced by gut microbes, drugs such as acetaminophen (the main ingredient in Tylenol) and statins, as well as a variety of fat-soluble pesticides, food additives, flavor enhancers, good old-fashioned alcohol, and volatile organic compounds.

If that's not enough to get your attention, consider this: The liver also controls blood levels of certain vitamins and hormones. It activates vitamin D and metabolizes vitamin A, and also activates thyroid hormone by converting T4 to the active version T3. It's also responsible for clearing T3 from the blood.

The human liver is really a magnificent organ. It is the only organ in our bodies that is able to regenerate. When other organ tissue, like the heart, is damaged it is replaced by scars. Unlike every other organ in the body, when the liver is damaged it can be replaced with new liver cells.[2] While prolonged exposure to liver toxins, the most famous of which is alcohol, can compromise the liver's ability to regenerate itself, it is remarkably resilient. Chances are you know someone who drank to excess and lived a healthy life well into old age. I certainly do.

Unfortunately, the liver is also one of the organs most vulnerable to glyphosate. A large number of studies, spanning decades, have demonstrated that glyphosate is markedly toxic to the liver, and that the liver is one of the first organs to be affected by glyphosate exposure. This is perhaps not surprising, since it is the liver's responsibility to clear toxins and toxic chemicals from the blood. As such, its exposure to glyphosate is higher than that of other organs. We know that glyphosate harms the liver in many different species, including fish, lizards, rabbits, and rodents.[3] We also know it harms the human liver in a variety of ways. An extensive body of scientific research reveals that glyphosate disrupts cytochrome P450 (CYP) enzymes, depletes the liver of glutathione, induces oxidative stress (which can then lead to mitochondrial impairment and mitochondrial DNA damage), and causes fatty liver disease.[4]

Compounding the problem is that glyphosate is only one of many chemicals that puts stress on our livers. Toxic environmental exposures work synergistically to cause or exacerbate liver disease.[5] We are exposed to many herbicides, insecticides, and environmental pollutants every day. Many people also take a wide range of prescription and over-the-counter medications, and many of these medications are known to be toxic to the liver. Nothing good comes of combining glyphosate with liver-damaging drugs. In fact, glyphosate increases the cytotoxicity of other substances, resulting in synergistic damage to the liver. In some cases, the damage is so substantial that a transplant becomes the only viable option.

A Bit of Biochemistry

Proteins are assembled from amino acids like beads on a string according to a four-letter genetic code (AGCT, standing for the nucleotides adenine, guanine, cytosine, and thymine) dictated by the DNA. All cells that have a nucleus have the code to make all the proteins the body uses. But many codes are never expressed. When we say that a particular cell type "expresses" a certain protein, we're saying that it is one of the cell types that makes that particular protein from the code.

Some proteins are expressed constitutionally in a particular cell type, meaning that these cells always synthesize this protein at a certain rate regardless of circumstances. For example, eNOS is an enzyme that is constitutionally expressed in endothelial cells—the cells that line the interior of blood vessels. Its close cousin, iNOS (inducible NOS), is expressed in immune cells, but only when induced by certain circumstances, such as a bacterial infection. As we talked about in chapter 5, another term used by biologists is "upregulated" (and its counterpart "downregulated"). A protein is upregulated in response to certain signaling molecules, which are often cytokines. This means that the cell increases the rate at which it makes that protein. The rate of synthesis of a given protein might increase a thousandfold in response to a signaling molecule.

The liver produces many proteins with various functions, but a few are especially important. Doctors routinely measure the levels of some of these proteins in the blood and use the results as an indicator that the patient may be suffering from liver disease. In particular, high serum levels of alanine aminotransferase (ALT) and aspartate aminotransferase (AST) are used as metrics of liver disease. Another less often used metric is γ-glutamyl transpeptidase (GGT), an enzyme that breaks down glutathione. Higher than normal levels of GGT are associated with liver disease, as well as with diabetes, obesity, and cancer.[6] Then there's malondialdehyde, which is a nonprotein indicator of oxidative stress. The levels of AST, ALT, GGT, and malondialdehyde are regularly measured in laboratory, animal, and human studies when researchers want to assess whether a chemical might be toxic.

Progressive Liver Disease

Your doctor may have told you at some point that drinking too much alcohol can cause fatty liver disease. Nonalcoholic fatty liver disease (NAFLD), on the other hand, is a liver disease that is not caused by alcohol. It has become a global epidemic, and it is the most common cause of chronic liver problems worldwide.[7] In the United States alone, an estimated 80 to 100 million Americans have been diagnosed with NAFLD. That's equivalent to the entire population of Vietnam.

NAFLD is characterized by the accumulation of fat in the liver (steatosis), along with liver inflammation. This leads to fluid accumulation and scarring, and is a precursor to even more serious problems: hardening of the liver, liver cancer, and liver failure. One study of 156 obese children and young adults, ages 5 to 20, found that 19 percent were suffering from NAFLD. It's the most common form of liver disease in children, and incidents of nonalcoholic fatty liver disease have more than doubled in the last two decades.[8] People of all ages suffer from it. Obesity is a risk factor, although thin people are being increasingly diagnosed, as well.

NAFLD can be asymptomatic, at least initially. But it can also lead to something called nonalcoholic steatohepatitis, or NASH. Cases of NASH are increasing exponentially in the United States.[9] In 2002, NASH was the primary reason for liver transplants for 3 percent of patients. But in 2011, it was the primary reason in 19 percent of patients, a *sixfold increase* in less than a decade.[10] NASH is now the second most common reason for liver transplants after hepatitis C.

NASH is a progressive liver disease, and there is no effective treatment for it. It begins with excess accumulation of fats, often associated with oxidative stress. NASH primes the liver for severe, sometimes lethal, scarring as well as for a severe liver cancer called hepatocellular carcinoma. It can lead to damage of cellular fats and mitochondria, and eventually to cell death and organ failure. Fewer than half the people diagnosed with hepatocellular carcinoma are still alive after two years.

NASH is significantly more common in people with brain disorders. When 50,000 people with autism were compared to a larger cohort of people without autism, those with autism were at nearly three times greater risk for developing both nonalcoholic fatty liver disease and nonalcoholic steatohepatitis.[11] Disturbances in liver function, like disturbances in the gut, can lead to disorders of the brain. When the liver can't keep toxins out of the blood, the brain is subjected to more exposure.

In a 2020 study, patients with biopsy-confirmed NASH had significantly higher levels of glyphosate residues in their urine than people without liver disease. Those with advanced disease had higher levels than those in an earlier stage. And patients with the most severe scarring in their livers, a condition called advanced fibrosis, had significantly higher levels of glyphosate than the patients with less severe liver disease.[12]

Lousy Livers

One of the first organs to be damaged by glyphosate, the liver can be damaged at levels of exposure considered environmentally "safe."[13] Liver damage, along with kidney damage, was one of the most pronounced effects in the long-term rat study conducted by Séralini and his team.[14] Glyphosate levels in the kidneys and livers of farm animals are approximately a hundredfold greater than glyphosate levels found in fat or muscle.[15]

Dr. Ted Dupmeier, a veterinarian in Saskatchewan, Canada, argues that feeding livestock glyphosate-contaminated genetically modified corn severely compromises their health. A postmortem examination of the livers of cows revealed "fat, large livers that were mottled and friable, or like sawdust inside." After removing feed that had been sprayed with glyphosate the herd became much healthier.[16]

In a 2019 study, Pakistani scientists looked at the metabolic effects of three levels of glyphosate exposure on rabbits.[17] Exposed rabbits showed negative effects that increased over time, and the higher doses of herbicide resulted in the worst outcomes. That is, there was both a time-related and a dose-related correlation with metabolic dysfunction. The rabbits' white blood cells and platelet counts went up, while their red blood cells and hematocrit went down. (Hematocrit is the percent of red blood cells by volume in the blood.) The red blood cells lost significant amounts of hemoglobin, which was converted into bilirubin. A marker of hemoglobin breakdown, the presence of bilirubin demonstrated that the red blood cells were damaged by glyphosate. At the same time, other standard markers of liver disease increased, including serum ALT, AST, alkaline phosphatase, urea, and creatinine. These are all indicators of liver and kidney damage.

These results are consistent with other animal studies. Albino mice exposed to a single dose of glyphosate and then sacrificed 72 hours later showed classic metabolic markers indicative of liver and kidney disease.[18] In these mice, as in the rabbits, glyphosate induced a significant increase in serum AST, ALT, urea, and creatinine. Glutathione levels in the mice were depleted, an indicator of oxidative stress, and malondialdehyde levels were elevated in the liver and kidney tissues. Higher frequencies of chromosomal aberrations and micronuclei in the tissue samples confirmed oxidative damage to DNA, a risk factor for cancer.

In 2018, Chinese researchers looked at effects to the livers of mouse offspring whose mothers had been exposed to glyphosate during pregnancy.[19] This study included measures of blood parameters related to fats and cholesterol, as well as standard liver enzymes. The offspring did not do well. The scientists found elevations in the serum of both AST and ALT as well as enhanced liver expression of HMG coenzyme A (HMG-CoA) reductase, the rate-limiting enzyme in cholesterol synthesis. The offspring also had significantly elevated levels of serum total cholesterol, LDL, and triglycerides. These findings are important. They suggest that glyphosate may be a primary cause of elevated serum cholesterol in humans.

Even when rats are administered Roundup via drinking water at doses so small they are deemed safe, their livers suffer. In fact, a detailed study of protein expression in the damaged livers revealed characteristic features of severe liver disease.[20] If we see these effects when animals are given very low doses, what happens when they receive high doses? In 2018, scientists in Egypt looked at the high-dose effects on albino rats, administering 25, 50, and 100 milligrams per kilogram of body weight of Roundup to the rats every day for 15 days. The effects on the liver were catastrophic: "increase of enzymes activities of ALT and AST, cellular infiltration, many signs of nucleus degeneration, focal necrosis, rarified cytoplasm, disorganization of cellular organelles, and deposition of lipid droplets." [21] These are truly devastating morphological defects, and this study shows severe disruption of healthy tissue courtesy of glyphosate.

In 2019, scientists in Tunisia replicated these results. They examined the effect of high-dose glyphosate exposure to the liver function of rats and found significant evidence of oxidative stress. They specifically saw increases in malondialdehyde and damaged proteins and decreases in multiple antioxidants: superoxide dismutase, catalase, glutathione, and vitamin C all went down.[22]

Glorious Glutathione

One of the most important molecules we need to stay healthy and disease-free is one that most people, even many health care professionals, have never heard of: glutathione (which I told you about in chapter 5). This molecule is a master of detoxification and a major antioxidant. I like to think of it as nature's mop. Like a mop, glutathione combines with toxic molecules and ushers them out of the body. It also mops up free radicals that would otherwise damage tissues. Glutathione is almost universally found in living organisms. Bacteria, fungi, plants, animals, and humans all make use of it. Present in the liver in large amounts, glutathione is assembled from three amino acids: glutamate, cysteine, and glycine.

Glutathione protects liver cells against oxidation damage. The liver, as it goes about its business of detoxifying fat-soluble toxins and toxic chemicals, needs lots of glutathione to keep it safe. In the immediate presence of glyphosate, the body will make more glutathione. But in the long term, glutathione levels will fall and become depleted. Plants get similarly stressed, even though they don't have livers, of course.[23] Glyphosate causes plants, including corn, peas, peanuts, and wheat, to upregulate glutathione and glutathione S-transferase, an enzyme that conjugates glutathione to multiple toxic chemicals to make them more water soluble so that they can be cleared. It's unlikely that glyphosate itself gets conjugated to glutathione. It's more likely that glyphosate interferes with other detoxification channels, and as a result glutathione is upregulated. Even at low levels, glyphosate has been shown to induce oxidative stress. However, as I discuss below, glyphosate could also get *incorporated into* glutathione, by substituting for the glycine residue it contains.

When rats were exposed to glyphosate over a period of eight weeks, the glutathione in their blood, brain, heart, liver, and kidneys was greatly reduced.[24] At the same time, their malondialdehyde levels increased significantly. This suggests that, over time, the rats were unable to maintain adequate supplies of glutathione to counter the oxidative damage induced by glyphosate. In a 2012 study, scientists found that Roundup caused an elevation in GGT levels in the blood of exposed rats at 50 or 500 mg per kilogram of body weight over 15 days.[25] Remember that GGT breaks down glutathione. Plants consistently increase synthesis of glutathione in the presence of glyphosate, as do animals. But over time glutathione levels fall.

An explanation for this phenomenon is that glyphosate is substituting for the glycine in glutathione, preventing the glutathione from functioning properly. Glycine is the last amino acid in the tripeptide glutathione sequence. This would, theoretically, make it easy for glyphosate to substitute. Defective tripeptides would then need to be disassembled by GGT, for the possibility of a healthy reassembly without the glyphosate substitution. Glyphosate's substitution for glycine in glutathione synthesis would kick off a futile cycle of synthesis and degradation.

We talked about the importance of sulfur in chapter 6. As it so happens, glutathione is a stored form of sulfur. One of its three amino acids, cysteine, is a sulfur-containing amino acid and all three amino acids in glutathione are critical for sulfate synthesis. Glutamate and glycine are both needed to make heme, an essential cofactor for sulfate synthesis, and cysteine provides the sulfur atom.[26] So, in addition to its role as an antioxidant, glutathione is a reserve resource to supply critical nutrients needed to produce sulfate when sulfur levels in the blood become deficient. The idea that glyphosate is altering glutathione by incorporating into it in place of glycine could explain the impaired ability of animals exposed to glyphosate to handle oxidative stress and detoxify fat-soluble chemicals, as well as their inability to maintain adequate sulfate supplies in their blood vessels.

An Essential Enzyme

In chapter 3, I introduced you to phosphoenolpyruvate (PEP), one of the two molecules that come together to form EPSP synthase, the enzyme that glyphosate suppresses in the shikimate pathway. Now I'm going to introduce you to phosphoenolpyruvate carboxykinase (PEPCK), an enzyme that plays an essential role in gluconeogenesis by converting lactate, proteins, and fatty acids to glucose in the liver. As you will see, there are remarkable parallels between PEPCK and EPSP synthase.[27] These similarities offer insight into how glyphosate may be damaging the liver.

Children who are born with a rare genetic defect in PEPCK often have many debilitating health problems.[28] They often suffer from low blood sugar and acute episodes of severe exhaustion, muscle cramping, and body weakness due to the lactic acid buildup in the bloodstream. Complications from PEPCK defects begin early in life, starting with a failure to thrive in infancy and eventually leading to severe liver failure and kidney disease. People with this condition also experience muscle weakness, developmental delays, seizures, lethargy, microcephaly, and heart problems that can lead to heart failure.

I propose that glyphosate disrupts the function of the PEPCK in the liver. Remember that glyphosate suppresses EPSP synthase by interfering with PEP binding at a site where there is a highly conserved glycine residue and several positively charged amino acids nearby.[29] Glyphosate's suppression of PEPCK follows the same pattern. It interferes with PEP binding at a site where there is a highly conserved glycine residue and several positively charged amino acids nearby. This gets technical but I will walk you through it: The glycine residue at position 237 in the enzyme PEPCK binds to its substrate oxaloacetate in order to help secure it in place so that it can react with the other substrate, GTP. So PEPCK does two things: It takes a carboxyl group off of oxaloacetate and then adds phosphate (stolen from GTP) to it to produce PEP. It can also run in the reverse direction starting with PEP. In this case, the first step is to pry the phosphate loose from PEP, just like the first step in EPSP synthase's reaction. As in EPSP synthesis, the prying loose of the phosphate releases energy to fuel the reaction.

There are also two positively charged arginine residues (R87 and R405) that bind to the phosphate in PEP—remember that PEP stands for *phosphoenolpyruvate*—securing it in place through their nitrogen atoms. Those same arginine residues can be predicted to secure glyphosate's methylphosphonate as well, in the space where PEP should go. This suggests that glyphosate would be well suited to substitute for the glycine residue at location 237, clobbering the protein's ability to bind PEP and thus destroying its enzymatic activity. This phenomenon would affect PEPCK in the same way that it affects EPSP synthase.

It's also likely that glyphosate disrupts PEPCK by chelating manganese, a catalyst for PEPCK, making it unavailable. We know that glyphosate exposure severely depletes manganese in the serum of dairy cows.[30] It also disrupts the uptake of manganese in soy plants.[31] We also know that manganese deficiency induced in neonatal rat pups leads to a sharp reduction of PEPCK activity. In fact, these infant rats have only 60 percent of the catalytic activity of normal (that is, manganese-sufficient) rat pups. The resulting lower plasma glucose levels in the first 2 days of life are often deadly.[32]

Scientists who manipulate the genome of mice can help us understand the role that a particular protein plays in the body by virtue of its absence in genetically engineered rodents. Such an experiment was conducted by a group of university researchers who engineered mice to produce an inactive variant of cytoplasmic PEPCK. These mice developed almost normally in utero, but they suffered from a severe metabolic crisis after birth. They all died within 2 days.[33]

Looking for clues as to what went wrong, researchers carefully examined the livers and blood of the mice. Normally, cytoplasmic PEPCK is not expressed in the liver until birth. After birth, PEPCK is upregulated in preparation for handling digested foods and using them to satisfy metabolic needs. As you may recall, a crucial role for PEPCK is gluconeogenesis, which is the synthesis of glucose from metabolic intermediates of the citric acid cycle. These intermediates are small organic molecules such as malate and oxaloacetate. Like the children suffering from a genetic defect in PEPCK, the designer mice experienced a dramatic drop in blood sugar over their 48-hour lifespan. In that short time, a significant amount of fat had already accumulated in their livers.

This extensive lipid infiltration of the liver surprised the researchers, but it is the key to understanding what happened.[34] Among the metabolic disturbances found in these mice were sharply elevated levels of the amino acids alanine and aspartate in their blood, as well as elevated serum lactate and triglycerides. When PEPCK isn't functioning correctly, cells have a more difficult time using lactate and fat as a source of energy in the citric acid cycle, and this is why excess serum lactate and triglycerides are present in the blood.

This is consistent with what's been observed in animals exposed to glyphosate. Disruption of PEPCK can explain a lot of it. Fish exposed to levels of glyphosate-based herbicides commonly found in the environment show eerily similar effects as the genetically engineered mice.[35] These fish have elevated serum lactate and triglycerides, as well as elevations in the two enzymes that synthesize alanine and aspartate. When PEPCK is broken, oxaloacetate piles up and gets redirected toward aspartate synthesis. These fish also have reduced levels of glucose in their livers, muscles, and blood, consistent with impaired gluconeogenesis.

So, fish exposed to glyphosate-based formulations and mice with defective PEPCK have similar metabolic derailments. Disruption of PEPCK by glyphosate through substituting for its critical glycine residue could account for these observed effects. Both glyphosate and defective PEPCK also cause fatty liver disease. This further connects the dots.

When adult male rats are exposed to Roundup, they show similar markers of oxidative stress and liver damage.[36] Liver glycogen following exposure is depleted in a dose-dependent relationship. This is highly significant, because it can be explained by PEPCK deficiency. The liver can generate glucose when blood levels fall dangerously low through two different mechanisms: gluconeogenesis (which depends on PEPCK) or converting stored glycogen back into glucose. But when gluconeogenesis is disrupted, glycogen stores are used up to maintain an adequate level of blood sugar.

When a PEPCK gene is deleted in mice they show a dramatic drop in liver glycogen compared to normal mice.[37] Defective mice also have twice as high plasma fatty acids and a 10-fold increase in liver malate, one of the intermediates of the citric acid cycle. When a common freshwater fish known for its hardiness was exposed to glyphosate, it was found, among other things, that glyphosate suppressed activity of the enzyme in the citric acid cycle that converts malate to oxaloacetate.[38] This enzyme was probably suppressed because the oxaloacetate could not be processed to PEP by a broken PEPCK enzyme. The intermediates of the citric acid cycle, oxaloacetate and malate, pile up because their channeling into gluconeogenesis is blocked when PEPCK is defective.

While PEPCK is highly expressed in the liver and kidney, it is also expressed in other tissues. One fascinating study involved engineering designer mice to have PEPCK deficiency only in the liver.[39] The deficiency led to a buildup of intermediates in the citric acid cycle, which blocked their livers' ability to metabolize fats. The mice consequently developed fatty liver disease. It appears, in other words, that glyphosate's suppression of PEPCK may be the primary explanation for its association with fatty liver disease.

A genetic defect of the PEPCK enzyme can have dire consequences. Glyphosate's effect may not be as dramatic as a genetic mutation where all PEPCK molecules are affected, but it does explain how glyphosate could cause fatty liver disease. It also explains the strong association between the rise in type 2 diabetes and the rise in glyphosate usage on core crops. When PEPCK is defective, the set point for blood sugar becomes elevated, because there is a constant threat of blood sugar becoming dangerously low. Better to have an extra buffer of sugar in the blood at all times than to risk going into a coma when it suddenly drops. The liver can't rapidly increase sugar supplies through gluconeogenesis when muscle activity drives sugar down too rapidly. The constant supply of excess sugar in the blood causes glycation damage to blood proteins and lipids. These damaged molecules are collectively known as advanced glycation end products, and they induce an inflammatory response linked to many modern diseases, including diabetes, heart, kidney, and neurodegenerative diseases.[40]

Born to Run

PEPCK also plays an extraordinary role in muscle. Most of us lift weights or exercise so that we can be strong and improve our body image. But skeletal muscles, which collectively represent significant body mass, also play an important role in metabolism. Especially under conditions of intense exercise, our muscles consume enormous amounts of energy.

We've been taught that exercise is one of the most important things we can do to improve our long-term health. Longevity research shows that movement throughout the day is particularly beneficial. A fascinating experiment with a special group of designer mice seem to bear this out. Interested in learning more about the role PEPCK plays in muscles, researchers from Cleveland, Ohio, genetically engineered mice to overexpress PEPCK in their skeletal muscles by up to 10 times the amount of PEPCK that normal mice would make. The researchers had no idea what to expect. The paper they subsequently published expressed the astonishment they felt at their findings. It was called "Born to run; the story of the PEPCK-Cmus mouse." (PEPCK-Cmus refers to excess amounts of PEPCK expressed in the cytoplasm (C) in the muscles (mus).[41])

These mice, the researchers discovered, were extremely energetic! They were 7 to 10 times more active in their cages than the control mice, and they were able to run on the treadmill for long sessions without a break. They also ate almost twice as much as normal mouselings. But because of their extreme hyperactivity, they were also smaller and leaner. The designer mice also lived longer than the controls and were more fertile. They had a greatly increased number of mitochondria in their muscles. Their excess PEPCK pulled intermediates like oxaloacetate out of the citric acid cycle, which were then used via the intermediary PEP to produce glycerol. Glycerol, in turn, allowed the mice to synthesize triglycerides in their muscles that could fuel the mitochondria instead of sugar. Their muscles burned a tremendous amount of fat, which also protected them from obesity.

This study shows that PEPCK can be tremendously beneficial in the muscles. But what happens when muscles have *insufficient* PEPCK? As you might expect from the research on designer mice, PEPCK deficiency leads to a substantial decrease in activity and energy, a reduced number of muscle mitochondria, lower muscle mass, impaired ability of muscles to use fat as an energy source, and the accumulation of triglycerides in the blood, leading to their deposit into abdominal fat.

As I mentioned in chapter 3, sitting all day can be detrimental to your health. Our sedentary lifestyle, combined with a standard American diet (aptly nicknamed SAD), can cause us to accumulate excess abdominal fat. Americans are often accused of being slothful. We don't exercise enough, we drink too much beer, and we watch too much TV. It's no wonder we have an epidemic of obesity in the United States. It must be our fault. Right?

Not so fast.

We have also seen a sharp rise in chronic fatigue syndrome, a mysterious condition that baffles health care professionals. Chronic fatigue syndrome first emerged after glyphosate was introduced into the food chain. While experiments in laboratory animals produce clear and quantifiable effects, humans are admittedly more complicated. There are always a number of variables at play, so it's difficult, if not impossible, to assert that Cause X produces Effect Y in humans. But a growing body of scientific literature shows that glyphosate is disrupting PEPCK in the skeletal muscles, causing impaired utilization of lipids as a fuel source and a reduction in their mitochondrial supply, leaving the muscles impaired in their ability to perform sufficient exercise and a general sense of physical exhaustion. Is it possible that we're not so lazy after all, but that a combination of contaminated poor-quality food and overexposure to other persistent environmental poisons is what is making us fat and unhealthy? Is the slow kill, courtesy of glyphosate and other toxic chemicals, why so many of us feel so depleted so often?

Feeding Ractopamine to Livestock

The United States is the world's largest producer of beef, supplying 20 percent of the world market, and it is home to nearly 800 million acres of grazing and pasturelands, both public and privately owned.[42] But many people don't realize that most beef cattle spend only the first year of their short lives out in the open. After that, they are shipped to concentrated animal feeding operations (CAFOs), where they are fed processed GMO Roundup Ready corn, supplemented with soybean meal, wheat middlings, barley, cottonseed, and sugar beets. These foods are often highly contaminated with glyphosate.

There's another controversial practice common in the United States: Ranchers feed the drug ractopamine to livestock to accelerate their growth. They use this drug to increase the ratio of muscle to fat in the animals, largely in response to consumer demand for leaner meats; this demand having risen over the past 15 years.[43] Developed by Eli Lilly and Company, ractopamine is incorporated into pig feed in a product known as Paylean, fed to cattle in a product called Optaflexx, and fed to turkeys in a product known as Tomax. At least 160 nations have banned the use of ractopamine, including Russia, China, Taiwan, and all the members of the European Union. Why? Because it acts as an adrenaline analogue. It can cause increased heart rate, tremors, headaches, muscle spasms, and high blood pressure.[44]

Ractopamine stimulates the adrenaline receptors in the muscles, inducing a complex signaling response to alter muscle metabolism. Which proteins are disrupted in response to ractopamine? You guessed it: PEPCK. When more PEPCK is produced, the muscles are better able to metabolize fats.[45]

Ractopamine induces a metabolic profile in livestock analogous to the one induced in the born-to-run designer mice. It partially offsets the damage done by glyphosate by increasing the expression of an enzyme whose activity is suppressed by glyphosate. Without ractopamine, the meat from these glyphosate-exposed animals is likely to be fatty and unpalatable. But with it, it is likely to cause harm to humans who consume it.

A Clogged Filter

The liver is one of the most important organs in the body, and one of the organs most severely affected by glyphosate. All the metrics that we use to assess impaired liver function—elevated levels of serum AST, ALT, and GGT, low liver glutathione, high ratio of GSSG to GSH, and excess creatinine and urea—have been found to be associated with glyphosate exposure in animal studies. Humans with liver disease have higher levels of glyphosate in their urine than those with healthy livers. Glyphosate at doses below regulatory limits caused fatty liver disease in rats, and fatty liver disease is an epidemic among humans today, even among young people.

While I am not aware of any science that has explicitly examined whether glyphosate disrupts PEPCK, there is a strong argument that it does, based on analogies with its effects on EPSP synthase. Further support comes from the fact that PEPCK deficiency leads to the same disease profile that is observed with glyphosate exposure. If glyphosate does disrupt PEPCK, it would impact the muscles as well as the liver. This may partially explain the epidemic we see today in obesity and the rise in chronic fatigue disorders.

Think of the liver as the body's filtration system. Now imagine that filter clogged with junk and pitted with holes and tears. The human liver is astonishingly resilient. But it takes the lead in so many vital functions in our bodies that even a small amount of damage to the liver can have diverse and catastrophic consequences. When we slowly poison our liver, we compromise its ability to protect us.

— CHAPTER 8 —

Reproduction and Early Development

We have no treatments for improving sperm production in infertile men, and we have no idea about what is the cause of the condition.

—PROFESSOR RICHARD SHARPE, reproductive endocrinologist, The University of Edinburgh

If you're young and trying not to reproduce, getting pregnant might feel easy. But many women today delay pregnancy while they develop a career, and then find it a challenge to conceive when they start trying in their 30s. Conception is actually quite difficult. An astonishing number of biochemical processes must synchronize in order for an egg from a woman's body to leave her ovary, travel through the fallopian tube, and become fertilized by healthy sperm. Once the egg and sperm are joined they form a blastocyst. That blastocyst then must successfully implant into the lining of the woman's uterus, burrowing beneath the surface and starting its frenzied journey of cell division and growth. It quickly divides itself into cells that will work together to become a specialized organ that will nourish the baby, the placenta, and the cells that will grow into the actual baby.

The odds haven't always been quite so stacked against parenthood, however. In the United States, birth rates have been dropping steadily for more than a decade. The year 2018 had the lowest birth rate in 35 years. And while most conversations about birth rate and population tend to focus on political aspects—access to abortion, access to birth control, fewer teen pregnancies, a decline in marriage, young people's desire to establish careers before starting a family, even climate change, and GDP—that's not the entire story. Biological reasons also exist, and they're significant: One out of every seven couples of childbearing age worldwide is infertile; in about 30 to 50 percent of these cases, it's the male partner.[1] In 2017, nearly 2 percent of babies born in the United States were conceived with the help of assisted reproductive techniques.[2]

Many people find it hard to believe that, in addition to difficulty getting pregnant, complications during pregnancy are also widespread in the United States, and that our maternal mortality rates are among the highest in the industrialized world.[3] High blood pressure, heart problems, obesity, medical mistakes, and postpartum hemorrhage are some of the complications that lead to maternal death.[4] A sad fact: The maternal mortality rate in the United States more than doubled between 1987 and 2015. Yet, at the same time, maternal death rates declined in other developed nations.

It's not just that women are having trouble getting pregnant, staying pregnant, and surviving pregnancy and childbirth. We also have a high rate of premature births. Many newborns struggle to survive and thrive. The United States lags behind most other industrialized nations in child mortality, especially in infant mortality. In fact, the United States has the highest rate of death on the first day of life among industrialized nations.[5]

What all this tells us is that, while some people argue that pregnancy, childbirth, and life in general are much safer now than they were in the past, there are, in fact, abundant reasons for concern. Many of those reasons are uniquely modern: maternal age, weight, drug use (recreational drugs and pharmaceutical drugs such as the birth control pill), exposure to radiation and tobacco smoke, food, genetics, underlying health problems, and, most notably for our purposes, exposure to toxic chemicals. Exposure to such varied substances as fungicides, jet fuel, bisphenol A, DEET, and atrazine can have profound effects on germ cells during pregnancy.[6]

We should therefore be alarmed that glyphosate-based herbicides are also showing up in the bodies of pregnant women. When 71 pregnant women ranging in age from 18 to 39 years old were tested in central Indiana, glyphosate was detected in 93 percent of their urine samples.[7] The women provided two urine samples at different points during pregnancy, as well as a water sample from their homes. Women living in rural areas had higher levels of glyphosate than those living in urban areas. Could this be from direct exposure from neighboring farms? We know that glyphosate can pass through the placenta and into the baby's body. It's been found in the umbilical cord blood.[8] We also know it can also pass through breast milk, and it has been found in the serum of maternally exposed offspring.[9]

Hormones Matter

The endocrine system is a complex high-level regulatory system. Glands such as the adrenal glands, the pancreas, the thyroid gland, the sex glands, and the master pituitary gland release hormones that travel throughout the bloodstream as chemical messengers influencing your appetite, metabolism, moods, sleep, and more. A large number of toxic chemicals are endocrine disruptors. They either bind at the receptors to block signaling, falsely signal, or disrupt the synthesis or the transport of the hormones. Endocrine-disrupting chemicals, which can cause infertility, are often more detrimental at lower doses than higher doses, defying the commonly held notion that "the dose makes the poison."[10] Indeed, pregnancy complications, miscarriage, and birth defects tend to occur more frequently when pregnant woman are exposed to *lower* doses of endocrine disruptors. The timing of the exposure also matters.[11]

In 2003, scientists from Brazil discovered that glyphosate-based formulations such as Roundup disrupt the endocrine system.[12] These researchers exposed pregnant rats with relatively high doses of Roundup from day 6 to day 15 of gestation. (Rats are usually pregnant for about 21 to 23 days.) On day 21, the pups were removed via C-section. The baby rats were then examined for teratogenic effects, or birth defects. The researchers found that the lower the exposure level, the more pronounced the pups' health problems. Two percent of the pups in the low-dose group showed congenital deformities. Only 0.6 percent of the pups in the medium-dose group showed congenital deformities. And none of the pups in the high-exposure group did. The results raise the question: If the effects were this pronounced in the pups of dams exposed to the smallest amounts of glyphosate, what might happen to pregnant mammals exposed to even lower doses?

While the pups exposed to the lowest doses suffered the most, nearly all the babies with any exposure had retarded skeletal development as well as anasarca, a severe and generalized form of edema with widespread tissue swelling beneath the skin, usually related to liver, kidney, or heart failure. Meanwhile, the pregnant dams in the group with the highest exposure died within three weeks.

A review paper published in 2020 examined a large body of peer-reviewed research on glyphosate, with a specific focus on looking for evidence that glyphosate is an endocrine disruptor. It found that glyphosate exhibited 8 out of 10 key characteristics of an endocrine disruptor.[13] Among other effects, glyphosate has been found to disrupt thyroid hormone regulation; to suppress testosterone synthesis; to inhibit a critical developmental enzyme, known as aromatase, that converts testosterone to estrogen; and to act as the estrogen receptor to enhance estrogenic signaling in breast cancer cells.

Testosterone Troubles

How toxic is glyphosate to sperm? In 2018, researchers in Greece exposed human sperm in vitro to a relatively low concentration of 1 mg/L of Roundup. They found that Roundup had a deleterious effect on sperm motility associated with mitochondrial dysfunction.[14]

Atrazine is another herbicide with a long and ugly past. Invented and manufactured by the agrochemical company Syngenta, atrazine has been linked to complete feminization of male frogs through chemical castration.[15] In fact, most researchers consider atrazine to be more dangerous than glyphosate. In 2015, a group of Nigerian scientists studied glyphosate's safety profile compared to atrazine and found something surprising. Testosterone levels, sperm motility, sperm counts, live/dead ratio, and the weight of the epididymis (the duct behind the testis that passes sperm to the vas deferens) were all *lower* in male rats exposed to glyphosate compared to those exposed to atrazine. Exposure to both chemicals simultaneously produced even worse effects on sperm, testosterone synthesis, and male reproductive organs.[16]

How do the male pups of pregnant dams exposed to glyphosate fare? Not well. In a 2019 study, female rats were given glyphosate in their drinking water at three different concentrations: 0.5 milligram per kilogram of body weight per day; 5 milligrams per kilogram of body weight per day; and 50 milligrams per kilogram of body weight per day.[17] All the rats exposed to glyphosate gave birth to male offspring with fertility problems. Even at the two lower levels of exposure, the male offspring had visibly deformed testes and low testosterone. Perhaps most surprising was that the pups of the dams exposed to the lowest amount of glyphosate—0.5 mg/kg/day—had an *89 percent decrease* in their sperm count.[18] While some scientists argue that it is the added ingredients, the adjuvants and surfactants, in glyphosate formulations that damage the sex organs and endocrine system, this particular study is significant because it shows without a doubt that glyphosate itself can cause reproductive defects and abnormalities.

Also in 2019, an international team of researchers led by Italian scientists exposed pregnant rat dams to Roundup via drinking water throughout gestation and for 13 weeks after weaning. The rats were given daily dosages comparable to those considered "safe" for humans, which according to the US acceptable daily intake (ADI) is 1.75 milligrams per kilogram of body weight per day. But the Roundup was anything but safe. Among other pathologies, female offspring whose moms were exposed to Roundup had abnormally high levels of testosterone.[19]

Yet another 2019 study examined what happens to mice and their offspring when they are exposed to glyphosate during gestation and lactation. The scientists used a concentration of 0.5 percent Roundup in the mice's drinking water and, again, the male offspring suffered serious health effects: delayed testicular descent, reduced sperm counts, and structural alterations inside their testes. These young mice also had increased concentrations of luteinizing hormone, which stimulates the production of testosterone from Leydig cells, causing an increase in testosterone in their testicles.[20] The mice were likely making more testosterone to compensate for their bodies' reduced ability to make sperm. At least five other studies from the United States and Argentina, all published in 2020, confirm these findings: Glyphosate-based herbicides are endocrine disruptors that decrease fertility when offspring are exposed during critical moments of their development.[21]

Disrupting Leydig Cells

Leydig cells are testosterone-producing cells in the testes that have large round nuclei. Roundup inhibits hormone production in these cells primarily by disrupting a protein called steroidogenic acute regulatory protein, or StAR, that regulates testosterone synthesis.[22] Roundup exposure at a level that is not toxic to Leydig cells themselves can block hormone production by 94 percent and reduce the activity of the regulating protein by 90 percent.

This protein controls the rate-limiting step in hormone synthesis in both the adrenal glands and the testes. The amount of regulatory protein synthesized by Leydig cells does not change in the presence of glyphosate, but the protein's enzymatic activity is suppressed.

Why is this happening? Glyphosate and glyphosate-based herbicides likely suppress PEPCK activity in the testes. PEPCK, important for protecting the body from fatty liver disease, diabetes, oxalate toxicity, and neonatal hypoglycemia, is also essential to activate steroid synthesis, including sex hormones.[23] PEPCK increases mitochondrial activity to generate ATP, and Leydig cells depend on ATP to modify the regulatory protein StAR.[24] With insufficient ATP due to suppressed PEPCK activity, the enzyme remains quiescent and testosterone cannot be synthesized. We know that healthy cells are more sensitive to mitochondrial disruption than cancer cells, which rely on glycolysis outside of the mitochondria to supply their energy needs. Because StAR controls cholesterol transport into the mitochondria where the enzymatic conversion of cholesterol to testosterone takes place, this research also suggests that glyphosate is playing a role in disrupting mitochondria. Might this be a reason for the seemingly inexplicable and incurable mitochondrial dysfunction we are seeing in some children?

The enzyme PEPCK is critical for synthesizing testosterone, which is derived from cholesterol, in the testes.[25] Through an elegant set of experiments, a large group of researchers in South Korea confirmed that luteinizing hormone, which is released by the pituitary gland as puberty approaches, induces the expression of PEPCK in the testes, and that suppression of PEPCK interferes with testosterone synthesis. Isocaproic acid is a fatty acid and byproduct of the cholesterol metabolism that is needed to synthesize testosterone. In healthy cells, isocaproic acid is then channeled into the citric acid cycle to produce ATP, which is also needed to produce testosterone. However, if PEPCK is dysfunctional, intermediates do not get removed from the citric acid cycle, preventing the metabolism of fats.

Long QT syndrome is an electrical problem in the heart, causing it to take longer than normal to recharge between beats. It leads to an irregular heart rhythm, which can cause sudden cardiac arrest, and it's often the cause of death in people who commit suicide by drinking glyphosate-based herbicides.[26] One explanation may be the disruption of testosterone synthesis by the Leydig cells. Androgen-deprivation drugs, which are used to treat prostate cancer by suppressing the synthesis of testosterone, put cancer patients at substantially increased risk of having long QT syndrome.[27] Testosterone supplementation, on the other hand, has been shown to shorten that elongated interval, protecting the heart against this syndrome.[28]

String of Pearls

A woman's ovaries continuously grow egg-containing follicles, one of which gets released every month. If the eggs are not released during ovulation, however, the ovaries can grow enlarged and the lining of the uterus can become inflamed. If you see inflamed, fluid-filled cysts in the ovaries on an ultrasound image, they look a lot like a string of pearls.

Polycystic ovary syndrome (PCOS) is one of the most common causes of female infertility and endocrine disorders in women, although some argue that it's actually underdiagnosed.[29] It is associated with an accumulation of cysts (fluid-filled sacs), which are egg follicles that didn't get released during ovulation. This condition affects as many as 20 percent of women worldwide.[30] It's categorized by painful menstrual periods, excess body hair, acne, and dark patches on the skin. Three-quarters of women with PCOS struggle to get pregnant.[31] Some scientists believe that, because it disrupts the gut microbiome and induces leaky gut syndrome, glyphosate can contribute to PCOS.[32] Elevated levels of zonulin, the protein that modulates permeability of the digestive tract, are associated with PCOS.

Another characteristic of PCOS is an abnormally high level of male hormones such as androstenedione and testosterone. This characteristic is called hyperandrogenism. In addition to irregular menstrual cycles and hirsutism (a male pattern of hair growth), women with PCOS often suffer from classic features of metabolic syndrome: obesity, insulin resistance, and an increased risk of developing type 2 diabetes and cardiovascular disease.[33] Women with PCOS also have a higher risk of having children with autism.[34]

Researchers have struggled to find genetic links to PCOS, and for the most part they have struck out. Most women with PCOS don't have any obvious genetic mutation to account for their disease. At the same time, an abundance of scientific evidence suggests that PCOS is environmentally induced. Suppressed activity of at least three different enzymes involved in steroid metabolism have been found in association with PCOS: hexose-6-phosphate dehydrogenase (H6PD),[35] aromatase (a CYP enzyme),[36] and PAPS synthase (the enzyme that activates sulfate).[37] As we've discussed in previous chapters, all three enzymes are disrupted by glyphosate through glycine substitution.

Glyphosate's disruption of aromatase further imbalances sex hormone production. Aromatase is the key enzyme responsible for converting testosterone to estrogen. When aromatase can't turn testosterone into estrogen, testosterone builds up and estrogen becomes deficient. Glyphosate's suppressive effects on aromatase have been directly demonstrated. Human placental cells exposed to Roundup at concentrations a hundred times *lower* than the recommended use in agriculture show disruption of the activity of aromatase.[38] This dovetails with experiments that show that glyphosate disrupts CYP enzymes in rat liver, since aromatase is a CYP enzyme.

Sex hormone binding protein (SHBP) is a group of proteins (a.k.a. a globulin) that is produced in the liver and circulates in the blood. As the name suggests, SHBP binds to the sex hormones, reducing their bioavailability. Androgens ("male" hormones) decrease expression of SHBP, making sex hormones more bioactive. Fat tissue, while once viewed as basically inert, is actually an endocrine organ, and fat cells have the capacity to produce sex hormones. When synthesis of DHEA sulfate is blocked by glyphosate, fat cells produce more sex hormones, and, when aromatase is blocked, they produce more androgens relative to estrogen. These endocrine disruptions cause increased bioavailability of the overproduced androgens, leading to PCOS. Low levels of SHBP are also a strong predictor of type 2 diabetes in both women and men.[39]

A woman with both abundant fat tissue and who is chronically exposed to glyphosate has a big problem. Glyphosate interferes with DHEA sulfation, and this causes the fat cells to make more sex hormones. The fat cells make more androgens relative to estrogens because glyphosate blocks aromatase. Androgen overexpression decreases the supply of SHBP, which then increases the activity of the androgens. This is a direct path to polycystic ovary syndrome. Low SHBP is a marker for diabetes, and obesity is a risk factor.

DHEA Sulfate and Anencephaly

You may remember from chapter 6 that DHEA sulfate is severely depressed in association with a developmental disorder called anencephaly, where a child is born without a cerebral cortex. Dr. Gregory Nigh, an oncologist based in Portland, Oregon, and I published a paper in 2017 explaining the myriad ways, including through disruption of DHEA sulfate, that glyphosate could be expected to cause anencephaly and other neural tube defects.[40] Defects in neural tube closure occur in 1 in 1,000 births, making this the second most common birth defect.[41]

There was a bizarre epidemic of anencephaly in Yakima, Benton, and Franklin Counties in Washington State during 2012–2013. Researchers investigated many factors to identify what the victims had in common, but they struck out and were baffled.[42] They did not think to investigate a glyphosate link, because glyphosate is presumed to be nontoxic to humans. However, when Dr. Don Huber investigated the situation, he discovered that, in 2010, 2011, and 2012, significant amounts of glyphosate in the formulation Rodeo were dumped into the three local rivers to control invasive aquatic weeds. These rivers provide irrigation and drinking water to these three counties. Within two years, farm workers began to experience an exceptionally high rate of anencephaly and miscarriage.[43]

Vitamin A and Methylation Pathways

Vitamin A plays a critical role in embryonic development. Although it's essential for the proper development of a fetus in the womb, too much vitamin A can be harmful to both the mom and her unborn baby. Vitamin A is expressed at critical developmental stages and then quickly cleared by specialized enzymes. Several different CYP enzymes are involved in metabolizing vitamin A.[44] One is essential for embryonic development and another is essential for both postnatal survival and germ cell development.

In African clawed frog embryos as well as in chicken embryos, glyphosate exposure, whether as pure glyphosate or as part of a formulation, results in impaired neural tube development, leading to teratogenic effects such as microcephaly and cyclopia (a single eye in the center of the head).[45] The reason for these grotesque deformities seems to be an increase in activity of retinoic acid, a metabolite of vitamin A, in exposed embryos. In fact, teratogenic effects can be reversed by cotreatment with a retinoic acid antagonist.

What do African clawed frogs and chickens have to do with humans? Hospital records for 21,844 newborns in Argentina from over 850,000 births between 1994 and 2007 revealed a concentration of birth defects in the central region of Córdoba province, an agricultural area where maize, soy, and wheat predominate and where glyphosate is heavily used on crops.[46] Indeed, scientists have found significantly higher incidences of multiple congenital disorders within the Córdoba province, including neural tube defects (spina bifida among them), microtia (underdevelopment of the outer ear), cleft lip, cleft palate, polycystic kidneys, polydactyly (extra fingers or toes), and Down syndrome. Many of these are associated with vitamin A toxicity.

Experiments on mice have demonstrated that exposure to high levels of dietary vitamin A could lead to neural tube defects. This is especially true when the mice were fed a diet high in folic acid and methyl donors such as methionine, betaine, and choline.[47] (Methyl donors are nutrients that carry methyl groups, and folic acid facilitates their transfer from one molecule to another.) This is ironic and surprising, because low folic acid is also linked to neural tube defects, which is why pregnant women are exhorted to take folic acid. The fact that low folic acid and high folic acid are both problematic suggests that it is absolutely critical for retinoic acid to be swiftly metabolized by CYP enzymes in order to precisely control the timing of various stages of development.

A Farmer in France

A case study of a farmer whose three children suffered from an unusual number of genetic defects reveals just how dangerous glyphosate can be. His children's health problems included defective anuses, stunted growth, hypospadias (a birth defect where the opening of the urethra is on the underside of the penis instead of on the tip), heart problems, and unusually small penises.[48] This farmer wore gloves while he hand-sprayed his crops with glyphosate but did not don a breathing mask or protective suit.

French toxicologists measured the amount of glyphosate in the farmer's urine and found that he had no detectable amount immediately prior to applying his crops with a glyphosate-based herbicide, but that the amount of glyphosate in his urine quickly peaked to a high of 9.5 micrograms per liter just 3 hours after he finished the herbicide application. This father's exposure was most likely from breathing it into his lungs and absorbing it through his skin. Glyphosate was also detected in the urine of one of his children, despite the fact that his family lived nearly a mile away from their fields. It's unclear how the child became exposed. The farmer may have tracked it home on his clothing. Or the wind may have carried it from the field.

Seven Generations

Glyphosate has powerful epigenetic, transgenerational effects, something scientists call generational toxicology.[49] In epigenetics, a specific methylation pattern can develop in the germ cells in utero, and this pattern can be passed on intact through multiple future generations. In fact, sometimes the most devastating effects often don't appear until subsequent generations.

A carefully designed 2019 study at Washington State showcases this. For this experiment, pregnant rats were exposed to very low doses of glyphosate between the eighth and fourteenth days of gestation only. The dosing was intentionally kept low, at half of the no observed adverse event level (NOAEL). (This term is used universally for toxic chemicals and it refers to the level below which animal experiments have been unable to observe harmful effects.) There was no observable harm to the pregnant rats or their offspring. While this should be a cause for celebration, there was a rub. The researchers found dramatic increases in several severe health problems, most linked to reproduction, in the second and third generations.[50]

The grandpups and great-grandpups of the exposed rats had damaged ovaries, mammary glands, testes, and prostate glands—health problems that could not be explained by anything else. The most common abnormalities observed in the testes were atrophy and vacuoles, along with sloughed cells and debris. The third-generation females had a 40 percent increase in kidney disease, such as fluid-filled cysts and shrunken glomeruli, which is where waste products are filtered in the kidney. A unique pathology rarely observed in previous studies of other chemicals was glyphosate's adverse effects on the birthing process in future generations. Thirty percent of the second-generation offspring either died during birth or gave birth to pups that died at birth. There was also an inexplicably high obesity rate in the third generation.

A similar study published a year earlier, in 2018, involved exposing pregnant rats to low doses of glyphosate beginning early in pregnancy and during childbirth.[51] Like the rats in the 2019 Washington State study, the exposed dams in the 2018 study had no obvious health issues. However, their first-generation descendants, although they appeared to develop normally, had significant issues during pregnancy: placental stress, retarded fetal growth, and premature birth. Remarkably, three offspring from three different mothers suffered from extraordinarily rare genetic deformities (conjoined twins, missing limbs), where, statistically, there should have been none.

How could it happen that a chemical that has no apparent effect on the pregnant dam has such profound effects on her offspring and her offspring's offspring? Especially when the offspring's exposure happens only in the womb? A female fetus produces all of her eggs very early in the gestational cycle. During the time when those eggs are being produced, they are extremely sensitive to environmental insults. If her offspring are born alive and manage to live into adulthood, insults from those exposures can lead to both genetic mutations and epigenetic effects that can survive through multiple future generations.

It's said that the Iroquois have a belief that all people are connected by a community that transcends time, from the very first people who walked the Earth to those who have not yet taken their first steps. With this view in mind, our job, as humans, is to bridge the gap between those who came before us and those who are to come. "The Peacemaker taught us about the Seven Generations," Oren R. Lyons Jr., a human rights activist and member of the Seneca tribe, explains. "He said, when you sit in council for the welfare of the people, you must not think of yourself or of your family, not even of your generation. He said, make your decisions on behalf of the seven generations coming, so that they may enjoy what you have today."[52]

Glyphosate's unique and diabolical effects do not just affect those of us who are alive today. Its toxic legacy will be passed down to the next generation and beyond. Even when glyphosate exposure shows no obvious harm to an animal during pregnancy and lactation, its progeny through multiple generations have been found to suffer from congenital defects, brain damage, endocrine disruption, and fatty liver disease.[53] Even if we stopped all production and use of glyphosate today, people may experience the fallout for generations to come.

— CHAPTER 9 —

Neurological Disorders

The scientific research is now abundantly clear: toxic chemicals are harming our children's brain development. As a society, we can eliminate or significantly lower these toxic chemical exposures and address inadequate regulatory systems that have allowed their proliferation. These steps can, in turn, reduce high rates of neurodevelopmental disorders.

—IRVA HERTZ-PICCIOTTO, PhD,
environmental epidemiologist

In 2018, a 66-year-old man attempted to commit suicide by drinking Roundup. In the hospital, he was treated for glyphosate toxicity. But his condition progressively worsened. He became disoriented, showed signs of amnesia, and started hallucinating. His doctors had already ruled out the presence of a tumor or an infection. Instead, he was diagnosed with acute toxic limbic encephalitis, an acute inflammatory response in the brain. After five months, this poor man was still suffering from amnesia, which was eventually diagnosed as permanent. His hippocampus, a curved part of the brain that is responsible for forming and storing memories and associated with learning and emotions, had shrunk by 23 percent.[1] To put this in perspective, a typical rate of shrinkage for Alzheimer's is 3.5 percent a year.

While most of us are not in danger of acute glyphosate poisoning, anyone approaching retirement age is bound to be anxious about the possibility of succumbing to a neurological disease such as Alzheimer's, Parkinson's, multiple sclerosis, or ALS. An estimated 10 percent of people over 65 and an estimated 50 percent of people over 85 in the United States suffer from

Alzheimer's.[2] In 2017, dementia was the primary cause of 261,914 deaths, up from 84,000 in 2000.[3] These CDC figures, based on death certificates, are likely an underestimate, because people live with dementia for many years before something more specific kills them. Some argue that we humans are just living longer, so it makes sense that our brains are giving out, but even adjusting for age, our dementia rates have more than doubled since 2000.[4]

As we discussed in chapter 4, Alzheimer's disease is the dominant subtype of dementia. It's characterized by problems with memory and executive function. A person suffering from late-stage Alzheimer's can forget such things as where they are, how to use a knife and fork, and how to go to the toilet alone. They often need constant care. The pharmaceutical industry has worked hard to come up with effective drugs to treat Alzheimer's, but there are no prescription medications that can arrest brain decline, let alone reverse it.

Scientists have started to recognize that Alzheimer's and autism have much in common. The diagnosis of autism encompasses a spectrum of disorders, from mild brain differences that some celebrate as neurodiversity to severe and devastating neurological damage. A complex disorder, autism is characterized by impairments in social interactions, a restricted repertoire of interests, stereotyped repetitive activities, and decreased cognitive ability. Between a quarter and a half of children with autism are nonverbal. Many will never live independent lives. The average life expectancy of a person with autism is only 36 years old.[5] Researchers at the University of Texas Health Science Center at San Antonio found that women with multiple chemical sensitivities have three times the risk of having a child with autism, as well as with other neurological disorders.[6] Children of these moms are also more prone to allergies. Multiple chemical sensitivities usually reflect excessive lifetime exposure to toxic chemicals.[7]

Norwegian researcher Olav Christophersen identified the accelerated rate of autism in the modern world as a canary in a coal mine. In 2012, when the rates were significantly lower, Christophersen argued that autism may pave the way to human extinction through a storm of mutations in our DNA caused by environmental toxins.[8] Cynthia Nevison, PhD, a Stanford-trained atmospheric scientist who works at the University of Colorado Boulder, examined whether the rise in autism was due to more prevalence or more diagnosis. She found that the increase in autism since the 1980s is not just the result of better diagnosis. Instead 75 to 80 percent of the increase is real.[9] This scientific analysis showed definitively that autism rates are rising. We can no longer claim that doctors are simply diagnosing it more.

At the same time, Nevison conducted a statistical analysis of 10 environmentally toxic substances most closely linked to autism to see how closely exposure to these substances mapped onto rising rates of autism among America's children. Her research revealed that lead, PCBs, organochlorine pesticides, car emissions, and air pollution all had *flat or declining* trends. That is, by 2014, when Nevison published this research, exposure had sharply decreased. Nevison identified only three toxic exposures that correlated strongly with autism: aluminum, polybrominated diphenyl ethers (PBDEs, which are compounds used as flame retardants), and glyphosate.

These are likely culprits. Aluminum is neurotoxic. When injected into the body as an adjuvant in a vaccine, it is engulfed by macrophages that transport it to stressed tissues, including the brain.[10] As for PBDEs, they disrupt the endocrine system. The association between exposure to PBDEs and impaired neurodevelopment has been studied extensively.[11] And then there's glyphosate. There is no question that glyphosate is contributing to the rise in autism, as we'll explore more in this chapter.

Toxic Combinations

The fact that exposures to other toxic chemicals may be contributing to brain disorders shouldn't diminish concern about glyphosate. These substances are working synergistically to harm the brain in even greater ways than they would independently. Aluminum is a good example. It binds to glyphosate. Two glyphosate molecules wrap around an aluminum ion to hide its +3 charge, making it easier for aluminum to cross barriers.[12]

A great deal of recent research implicates aluminum as a causal factor of brain inflammation characteristic of autism.[13] Aluminum has also been implicated in Alzheimer's disease and dementia.[14] Perhaps the best evidence comes from studying dialysis encephalopathy, a condition that has been linked to aluminum contamination in dialysis fluid for patients with kidney failure.[15] Dr. Christopher Exley, a professor of bioinorganic chemistry at Keele University in England, has also found high levels of aluminum in the postmortem brains of children with autism.[16]

In a different study, Exley's team analyzed aluminum distribution in the postmortem brain of a woman who had been exposed to high amounts of aluminum in the water supply over an extended period of time. The team found that the aluminum was concentrated within her glial cells and lymphocyte-like cells lining the choroid plexus.[17] The choroid plexus is embedded within the brain ventricles and is responsible for producing cerebrospinal fluid. An inflammatory condition, perhaps induced by the aluminum, had likely attracted the lymphocytes. This same pattern of intracellular, non-neuronal distribution of aluminum is also a feature of the autistic brains Exley and his team have studied.

The brain stem nuclei, including the pineal gland, are less protected from toxic metals than the cerebral cortex, because they lie outside the blood-brain barrier. A study examining concentrations of aluminum in various brain tissues found twice as much in the pineal gland as in any other part of the brain.[18] But it's also possible that glyphosate facilitates aluminum transport across the blood-brain barrier. As glyphosate chelates aluminum, it cancels out its +3 charge, making it easier for the aluminum-glyphosate complex to cross barriers.[19]

Conventional researchers have had an almost myopic focus on genetic underpinnings of autism. But it is becoming increasingly clear that the cause of brain impairment is a complex interaction between genetics and the environment.[20] Autism is associated with a long list of co-occurring health problems: gut dysbiosis, hormone imbalances, mineral deficiencies, heavy metal poisoning, metabolite disturbances, and inflammatory markers. In 1943, Dr. Leo Kanner, a psychiatrist at Johns Hopkins University, published a seminal paper that described autism for the first time based on his study of eleven children. Even at that time, when little was understood about autism and almost nobody had heard of it, Dr. Kanner described it as an illness with gastrointestinal disorders and dietary issues with behavioral manifestations.[21]

Analyzing data from over a thousand studies, James Lyons-Weiler, PhD, developed a theory that autism risk is primarily due to genetic predisposition, especially mutations that lead to loss of function in several important biological pathways of development or detoxification, interacting with environmental stressors.[22] The more of these predisposition-exposure combinations a person has, the greater the risk of brain dysfunction.

The largest and most recent investigation of pesticides and autism was conducted in the agricultural region of California's San Joaquin Valley.[23] Examining pesticides and herbicides, including chlorpyrifos and glyphosate, researchers found that pregnant women living within 1.2 miles of spray areas were 30 percent more likely to have children with severe autism. An odds ratio is a measure of the contribution of a particular factor to a disease. A high ratio means a stronger effect. When multiple chemicals are involved, it is computationally possible to tease out the effects of each chemical relative to the effects of the others. After correcting for other pesticides applied simultaneously, many of the elevated odds ratios became less pronounced in this study, but glyphosate's odds ratio *increased* once co-occurrence with other pesticides was accounted for. The highest odds ratio was found for children exposed to glyphosate during the first year of life.

How Glyphosate Harms the Brain

How could glyphosate be disrupting brain development? In several pernicious ways: by causing disruption in crucial enzymes; causing glutamate excitotoxicity; inducing yeast overgrowth; inducing gut dysbiosis; inducing cobalamin deficiency and sulfate deficiency; provoking imbalances in short chain fatty acids; provoking mitochondrial damage; and provoking thyroid hormone deficiency, to name a few.

Glyphosate can be taken up by cells along glutamate transport channels because, like the amino acid glutamate, it is negatively charged. This means it can take advantage of existing amino acid transporters to gain access across cell barriers that would otherwise be impermeable. For example, you know that epithelial cells on the skin, organs, and urinary tract serve as a barrier and are one mechanism of protection against invaders. But not against glyphosate. Some epithelial cells grown in culture actively take up glyphosate. These include epithelial cells that line the nasal cavity, the respiratory mucosa in the lungs, and the gastrointestinal tract.[24] This suggests that glyphosate may be able to make its way into the brain via the nasal passages and that glyphosate absorbed through the skin or ingested in contaminated food might also make its way into the brain. Normally, a small water-soluble molecule such as glyphosate would not be able to penetrate the blood-brain barrier, but one study has shown that neurons grown in culture with cells that make up the blood-brain barrier are vulnerable to glyphosate. Glyphosate seems to disrupt barrier function and permeate the brain via amino acid transporters.[25]

One area of the brain especially susceptible to damage from disrupted barrier function is the limbic system. The brain's limbic system is mainly concerned with emotions, memory, personality, appetite, social, and sexual behavior. This region consists of the thalamus, the hypothalamus, the amygdalae, and the hippocampus. Recall the 66-year-old man who attempted suicide by drinking Roundup and the dramatic shrinkage of his hippocampus that resulted. This tragic example is one of many that shows how the hippocampus, in particular, is sensitive to glyphosate. Glyphosate induces neurotoxicity in the hippocampus by overstimulating NMDA (N-methyl-d-aspartate) receptors.[26] Overstimulated NMDA receptors cause the cell to take up too much of the amino acid and neurotransmitter glutamate. Glutamate is an excitotoxin: It can cause neurons to fire at such a rapid rate that they become exhausted and damaged. Glyphosate can hitch a ride on the glutamate transporters and also enter the cell. As a comprehensive 2019 review explains, it's not just the glyphosate-induced gut dysbiosis that's harming the brain, inducing depression, and fomenting anxiety. It's also that glyphosate likely damages the hippocampus directly through glutamate excitotoxicity.[27]

The hippocampus is one of only two sites where new neurons can mature from stem cells and migrate into other parts of the brain. The other site is the subventricular zone, an area adjacent to the brain ventricles. When scientists investigated the effects of glyphosate on mouse neural stem cells, they found that glyphosate significantly reduced cell migration and differentiation, suppressed activity of a CYP enzyme that is known to be cytoprotective, and induced markers of oxidative stress. This work, published in 2020, suggests that glyphosate interferes with the maturation of new neurons in the brain.[28]

One mechanism for how glyphosate causes neurological damage at the cellular level is suggested in an experiment involving PC12 cells from rats. PC12 cells are derived from the adrenal medulla, which is the inner part of the adrenal gland. These PC12 cells have many neuron-like properties. They grow well in culture and have become a popular tool to investigate effects of different stressors on neurons. When scientists exposed PC12 cells to glyphosate and examined the consequences, they found that glyphosate decreased survivability of the cells in a dose-response way. It induced autophagy, which is a clean-up process to clear cellular debris. This suggests that glyphosate caused the buildup of damaged cellular components. Glyphosate also induced apoptosis, a form of programmed cellular death.[29]

Sulfate Deficiency and the Brain

Neurons, like all other cells, clear their cellular debris. They depend upon sulfate to do this. Much of the pathology in both Alzheimer's disease and autism traces back to sulfate deficiency. As you may remember from chapter 6, one important sulfated molecule, heparan sulfate, facilitates almost all core activities of a cell.[30] Many signaling molecules, particularly growth factors, bind to heparan sulfate to initiate their signaling response. Cell and blood vessel growth are regulated by heparan sulfate. Heparan sulfate also enables cells to take up fats, cholesterol, and fat-soluble vitamins.

Heparan sulfate is extremely important to neural development. It modulates neurogenesis, axon guidance, and the maturation of synapses.[31] And, in postmortem studies, it has been found to be deficient in the ventricles of the brains of children with autism.[32] When the ability to synthesize heparan sulfate is disabled in mice, the mice exhibit the features of autism.[33] I believe that heparan sulfate deficiency in the brain is a core component of autism and Alzheimer's disease.

Harm to Our Neurotransmitters

Free glutamate is the most abundant excitatory neurotransmitter in the vertebrate nervous system, accounting for over 90 percent of the synaptic connections in the brain. The hippocampus, which was so damaged in the man who drank Roundup, is especially rich in glutamate receptors. These receptors are essential for learning and memory. The hippocampus depends more on glutamate signaling than do other parts of the brain.[34]

There is no question that glyphosate disrupts glutamate. In 2014, Italian researchers found that Roundup caused glutamate to become neurotoxic in the hippocampus of rat pups. The pups were exposed to a low daily dose of Roundup in utero via their mothers' exposure and during lactation via their mothers' milk. The pups' mothers were fed water containing Roundup and their total exposure each day was about 40 percent, less than half of what the effects might be of the no observed adverse effect level (NOAEL). The pups were sacrificed at 15 days old and then their brain tissue was analyzed. In another branch of the same experiment, hippocampal slices were prepared from unexposed 15-day-old pups, and then those slices were exposed to Roundup at varying concentrations. The researchers found that exposure to Roundup increased calcium uptake in the cells by activating glutamate-sensitive NMDA receptors as well as calcium channels. Roundup increased the amount of glutamate released into the synapse by neurons. It also interfered with the ability of brain cells to clear glutamate from the synapses (by converting glutamate to glutamine).[35] Oxidative stress is an overabundance of free radicals that the body cannot clear. Excess glutamate in the synapse led to neuronal burnout, with oxidative stress damaging delicate components of the brain cells.

We've known since 2006 that children with autism have higher concentrations of glutamate in their brains compared to healthy controls.[36] They also have excess glutamate in their blood. A study comparing the plasma of 23 boys with high-functioning autism with the plasma of 22 healthy controls found that the boys with autism had statistically significant higher levels of glutamate in their blood (and lower levels of glutamine).[37] None of their other amino acids were out of line with the control group.

Glutamate Toxicity

To understand how glutamate can become neurotoxic, we need to understand how it operates in the brain. As a neurotransmitter, glutamate is a powerful stimulator of nerve cell activity. Normally it is sequestered inside nerve cell vesicles. It's released into the synapse in order to transmit a signal to a neighboring cell. It's then quickly cleared by surrounding astrocytes, so named because they are star-shaped brain cells. The astrocytes take up the glutamate and convert it to glutamine, a benign molecule without any neuroexcitatory effects, with the help of the enzyme glutamine synthetase. The glutamine is then shipped back to the neurons through a pathway that doesn't involve the synapse. The neurons take it up, convert it back to glutamate, and sequester it back in the vesicles.

The trouble is that the enzyme glutamine synthetase depends on manganese. And manganese can be chelated by glyphosate, making it unavailable.

When astrocytes lose their ability to convert glutamate to glutamine, the cycle is derailed. Glutamine synthetase also binds ATP as a source of energy. It contains a highly conserved string of amino acids, GAGCHTNFS, at the ATP binding site. Note that the second glycine (G) has an alanine (A) residue to the left, leaving lots of room for glyphosate's side chain. It also has a histidine (H) one away to the right, providing positive charge to attract glyphosate's phosphonate unit. You may recognize this as the glyphosate susceptibility motif I described in chapter 5.

I believe that glyphosate substitutes for this critical glycine residue and disrupts ATP binding of glutamine synthetase, leaving the astrocytes unable to clear glutamate from the synapse. Glyphosate not only causes glutamate to build up in the synapse, it also acts as a glycine analogue and binds with NMDA receptors.[38] And it gets taken up by the cell along glutamate transporters. Glutamate, together with glycine, excites NMDA receptors in the hippocampus, leading to neurotoxicity.

A follow-up experiment to the one that indirectly exposed rat pups to Roundup in utero and during lactation was performed three years later. This time researchers exposed rat pups to a similar dose of a glyphosate-based formulation (1% dilution in drinking water) both during gestation and after the pups were born. The baby rats were first exposed via their mothers, as in the previous experiment, and then drank the glyphosate-based herbicide directly via their own water. Their glyphosate exposure lasted until they were 2 months old. They were then put through a series of behavioral tests. As before, their hippocampal slices were then analyzed, this time when they were three months old.[39]

Before the pups were sacrificed, the exposed rats exhibited depressive behaviors. Afterward, researchers found glutamate disruption, NMDA activation, calcium uptake, and neuronal oxidative damage, just as in the previous study. This research, published in 2017, found that glyphosate also suppressed cholinesterase (an enzyme that breaks down acetylcholine, another neurotransmitter) in the hippocampus.

Another more recent study on mice has similarly found that chronic exposure to a glyphosate-based herbicide resulted in memory and cognitive impairments associated with reduced activity of cholinesterase and of antioxidant enzymes in the brain.[40] Glyphosate has also been found to significantly reduce cholinesterase activity in the brain of fish.[41] This is important because suppression of cholinesterase is a key component of the neurotoxicity of organophosphate insecticides. Glyphosate is an organophosph*onate*, which is not quite the same as an organophosphate. Still, it appears to disrupt acetylcholinesterase, leading to excessive cholinergic signaling that also contributes to neuronal toxicity.

Dietary Glutamate Causes Similar Effects

Katherine Reid, PhD, is a biochemist and the mother of a girl with autism. Dr. Reid saw improvements in her daughter's autistic symptoms when she put her on an organic diet and eliminated gluten and casein. Though the benefits of a gluten-free casein-free diet for people with autism is a subject of debate, many parents of children with autism find that their child's digestive symptoms improve when they eliminate gluten from the child's diet.[42]

Why would it help to eliminate gluten and casein? Katherine Reid realized that products made with gluten and casein are often high in glutamate. Most parents don't know that glutamate is a hidden ingredient in many processed foods. It's used as a flavor enhancer in the form of monosodium glutamate (MSG). MSG is so popular in Chinese cooking that it's even added to some brands of salt. It's harder to eliminate dietary glutamate than you might expect. Glutamate hides behind many disguised names on food product labels, including "hydrolyzed protein," "autolyzed plant protein," and "yeast extract." Reid started to eliminate glutamate from her child's diet. After she switched her family to an entirely whole food, real food diet and eliminated all processed foods, her daughter made a full recovery.

Excess glutamate is a known factor in several neurological disorders, including depression.[43] Abnormally high levels of glutamate lead to excessive oxidative stress in the brain, causing neuronal damage, particularly in the hippocampus. Food additives that contain excitotoxins such as monosodium glutamate and aspartame are "the taste that kills," according to retired neurosurgeon Russell Blaylock, MD.[44] To combat excess glutamate, Donald Rojas, PhD, cognitive neuroscientist and chair of the Department of Psychology at Colorado State University, has even proposed that pharmaceutical antagonists for glutamate receptors be repurposed to treat autism.[45] An

NMDA receptor antagonist, ketamine, has shown promise in treating both autism and depression.

So how is exposure to glyphosate affecting the brains of humans and other animals? There are three mechanisms for how glyphosate formulations and glyphosate itself could cause neuroexcitotoxicity in the hippocampus: by chelating manganese, making it unavailable as a catalyst for glutamine synthetase; by incorporating into glutamine synthetase in place of glycine at the ATP binding site; and by substituting for glycine as a neurotransmitter and overstimulating the NMDA receptors, causing the cell to take up excess glutamate. What's more, glyphosate also appears to suppress activity of acetylcholinesterase, operating in a way that is parallel to the neuronal damaging effects of organophosphate insecticides. While we still have a lot to learn about the exact mechanisms, it is clear from a growing body of scientific evidence that glyphosate harms the brain.

Sally's Story: The Role of B_{12}

Sally Pacholok worked as an emergency room nurse for more than 27 years. After decades of observing patients, she has come to realize that there is an epidemic of B_{12} deficiency in the United States. Vitamin B_{12}, also known as cobalamin, is a water-soluble vitamin that is critical for our red blood cells to form, our brains to function, and our DNA to synthesize. It's found naturally in animal products, including beef liver, clams, eggs, fish such as salmon and trout, milk, and poultry. A strictly plant-based diet contains no cobalamin.

It's not just that people aren't getting enough B_{12} from their food; people are also having difficulty making use of it. Cobalamin contains the mineral cobalt, and it can't function without its central cobalt ion, embedded in a corrin ring. The amount of two minerals, cobalt and manganese, in the blood can be extremely low in animals chronically exposed to glyphosate.[46] To make matters worse, the corrin ring (like the porphyrin ring in hemoglobin, which I described in chapter 5) is assembled from multiple pyrrole subunits, and a critical enzyme in the pyrrole synthesis pathway is inhibited by glyphosate.[47]

B_{12} deficiency can cause a host of health problems, including developmental delays, failure to thrive, reduced IQ, seizures, behavioral issues, and defects in fine motor control and gross motor movement. Untreated B_{12} deficiency in infancy can lead to permanent disability. Cobalamin is pathologically low in the brains of autistic children, the elderly, and in people suffering from schizophrenia.[48] In fact, levels of two forms of cobalamin, methylcobalamin (which is essential for methylation) and adenosylcobalamin, are up to three times lower in people with impaired brain function. Glutathionylcobalamin is even more dramatically reduced.[49] I find this fact particularly significant, as it suggests a glutathione deficiency is linked to brain disorders. Independently of cobalamin, low glutathione and impaired methylation are known features of autism. Both problems, compounded by cobalamin deficiency, negatively impact brain function.[50]

In addition to impaired methylation and low glutathione, low levels of vitamin B_{12} can ultimately result in high glutamate, high homocysteine, and a high amount of ammonia in the brain. These, in turn, induce an inflammatory response characterized as a low-grade chronic encephalopathy. We know that vitamin B_{12} inhibits the release of glutamate in neuronal synapses in a dose-dependent manner, so its deficiency would lead to an increase in the release of glutamate.[51] There is no question that low vitamin B_{12} is a compounding factor in glutamate toxicity linked to autism.

To their credit, Sally Pacholok and her colleague, emergency medical doctor Jeffrey Stuart, were able in some cases to reverse dementia—and even cure it—by treating patients with vitamin B_{12}.[52]

I became particularly interested in cobalamin's role in the body after learning about how glyphosate disrupts it. In 2019, I published another paper with Dr. Gregory Nigh, the oncologist based in Portland, in which we described how the body compensates for cobalamin deficiency. There are only a handful of enzymes that depend on B_{12}, but each is vitally important to cellular health. While these enzymes appear unrelated, they collaborate in an elegant dance to ameliorate damage from cobalamin deficiency. When one enzyme is disrupted, accumulated compounds will alter cellular metabolism to work around the disruption.[53] These accumulated compounds, namely propionate and methylmalonic acid, get into the brain and modify activities to extract sulfate from the brain in order to supply it to the macrophages and the blood vessels. This is essential to maintain healthy blood circulation and a healthy immune system, but it also causes the brain to become sulfate deficient.[54]

During brain formation, neuronal stem cells normally differentiate into either neurons or glial cells. Glial cells, which include oligodendrocytes, microglia, and astrocytes, form myelin to insulate nerve fibers and provide nutritional support and immune protection for neurons. This is all good. But when there are too many glial cells relative to neurons, the connectivity among neurons is disturbed. Glial cells are also responsible for initiating inflammation in the brain. Too many glial cells in the brain relative to neurons is a feature of autism.[55]

Propionate affects how cells differentiate. Too much propionate in the brain during early development will cause an overproduction of glial cells. An experiment on human neural stem cells demonstrated a dramatic effect of propionate on the differentiation into neurons or glial cells.[56] Under normal culture conditions, 47 percent of cells become glial cells, but when cells are exposed to propionate, a full 80 percent become glial cells. When propionate is injected into the brain ventricles of rats, they exhibit antisocial behaviors typical of autism. They also suffer from a neuroinflammatory response in the brain called glial cell activation, a condition associated with encephalopathy and autism.[57]

As early as 2002, autism was also observed to be associated with elevated methylmalonic acid in the urine, which could be the result of cobalamin deficiency.[58] Excess methylmalonic acid due to cobalamin deficiency inhibits a crucial enzyme in the mitochondria, the only enzyme that participates in both oxidative phosphorylation, which produces ATP, and the citric acid cycle, which metabolizes glucose.[59] The enzyme, a flavoprotein, is succinate dehydrogenase. Impairments in succinate dehydrogenase are, you guessed it, also associated with autism.[60]

Low levels of glyphosate-based formulations have been shown to induce cell death in human umbilical, embryonic, and placental cells, attributed to suppression of succinate dehydrogenase.[61] Since succinate dehydrogenase is a flavoprotein, it can be expected to be suppressed by glyphosate. As we talked about in chapter 5, flavoproteins bind the flavins FMN and FAD at a site where there are three highly conserved glycine residues, and these flavins contain phosphates that are displaced by the methylphosphonate unit in glyphosate.

Glyphosate Hates the Love Hormone

Oxytocin is often called the love hormone because it makes people feel good. This short neuropeptide is involved in inducing labor at the end of pregnancy and milk ejection from the mammary glands during breastfeeding, as well as in regulating complex social behaviors. It is a member of an ancient class of peptides dating back 600 million years.[62] Like other hormones in its class, oxytocin contains only eight amino acids, along with the nitrogen atom as a remnant of an original ninth glycine residue appended at the carboxyterminal (called G*). These peptides are all synthesized from a longer peptide. The final step in the synthesis of oxytocin is to cleave off most of the terminal glycine residue through enzymatic action, leaving just the nitrogen atom in place, which is labelled as G* to indicate its origins from glycine.

Glyphosate substitution for the terminal glycine likely disrupts this final step that splits up the glycine molecule. This is problematic because it is the glycine split that protects the peptide from degradation. And even if the cleft enzyme still works with glyphosate in place of glycine, it results in a version of the peptide with a methylphosphonyl group attached to its terminal end, which surely has significant, albeit unstudied, consequences on function. Earthworms, mollusks, insects, spiders, and vertebrates collectively synthesize at least 14 different short peptides like oxytocin. They are all made up of nine amino acids, ending with a snipped-off glycine residue. Glyphosate is likely disrupting their hormones through the same mechanism.

Oxytocin deficiency impairs social interactions. Children with autism are generally deficient in oxytocin. Treating them with supplemental oxytocin can lead to improvements in social interaction.[63] Could it be that their deficiency is due to glyphosate swapping in for the terminal glycine residue?

A Mother's Thyroid Matters

We know that many environmental pollutants, including ones used to control insects, weeds, and mold, are endocrine disruptors. The hypothalamic-pituitary-thyroid axis is one of the most affected neuroendocrine systems. A study involving over 35,000 people showed an increased risk of hypothyroidism in highly exposed individuals for several pesticides, including glyphosate.[64] Thyroid hormone is produced from tyrosine, one of the products of the shikimate pathway. Glyphosate's disruption of the gut microbiome likely suppresses the supply of tyrosine, leading to insufficient thyroid hormone production.

Two recent studies on rats exposed to glyphosate-based herbicides both in utero and during lactation found that their thyroid hormone processing was disrupted. In one study the offspring's genes expressed proteins as if the mothers suffered from hypothyroidism.[65] The other found adult male rats to have a significant increase in thyroid stimulating hormone, a hormone produced by the pituitary gland to induce more thyroid production. Early life exposure to glyphosate had altered their set point for thyroid-stimulating hormone.[66] While in the first study glyphosate levels were relatively high, this second study followed US guidelines for "safe" exposure. A plausible explanation is that, when the thyroid was deficient in tyrosine, it was stimulated to make more. Women with poorly functioning thyroids, whose bodies cannot make enough thyroid hormone, are nearly four times as likely to have a child with autism.[67]

Glyphosate's Synergistic Role in Widespread Neurological Dysfunction

Autism is not due to glyphosate exposure alone; that is clear. Autism, even the severe autism described by Leo Kanner in 1943, existed prior to the development of glyphosate. But it's equally clear that autism prevalence is rising, and dramatically so, as more toxic environmental exposures leach into our daily lives. Before the movie *Rain Man* came out in 1988, few people even knew what autism was. Today you'd be hard-pressed to find anyone who doesn't have a friend or family member with severe autism.

It is clear that environmental stressors are causing high rates of brain dysfunction, especially in children exposed during periods of rapid brain development and in older adults who have had chronic exposure over time. Some of these environmental exposures are likely causing evolutionary pressure, resulting in large numbers of mutations with unpredictable results. While it's possible that some of these mutations could be beneficial, many of them result in devastating dysfunction. Numerous environmental factors have been reliably tied to increased brain disorders.

One of the most insidious toxic exposures, affecting virtually everything we humans depend upon for maintenance of biological homeostasis—the food we eat, the air we breathe, the water we drink, the microbes that work so hard for us, the minerals we require, and even the critical proteins programmed by our DNA—is glyphosate. It's possible that in our increasingly toxic world, neurological dysfunction such as autism might increase even in the absence of glyphosate. But there is no question that the many ways in which glyphosate disrupts biological systems, especially the myriad ways it affects brain development, are contributing to the rapid accumulation of synergistic toxicities that we are witnessing. And that should scare us all.

— CHAPTER 10 —

Autoimmunity

> *Modern medicine asks* what *and* how: *what conditions do you have, and how do we treat them? But we should be asking why—this is the first critical step toward prevention. If we don't know why something happens, we can't hope to stop it.*
>
> —William Parker, PhD,
> Duke University School of Medicine

The United States spent nearly four trillion dollars on health care in 2019, far more per capita than any other country in the world. Health care is projected to account for nearly 20 percent of our gross domestic product by 2028.[1] At the same time, as I mentioned in chapter 8, our health outcomes lag behind most countries in the industrialized world.[2] And the Centers for Disease Control and Prevention (CDC) estimates that 90 percent of the nation's health care expenditures are for people with chronic and mental health conditions.[3]

The percentage of the population affected by autoimmune disease has been steadily rising in the United States. Today at least 41 million American are suffering from autoimmune disease.[4] What is an autoimmune disorder? It's when the body gets confused and treats its own proteins and DNA as if they were foreign molecules, causing diverse symptoms. It mistakes its own tissues as being a threat, often attacking the organs and joints. If you have an autoimmune disease, you have an overreactive adaptive immune system, an excessive antibody response to invasive pathogens that leads to the development of autoantibodies.

There are at least 80 different autoimmune diseases. Addison's disease, antiphospholipid syndrome, autoimmune thyroiditis, celiac disease, chronic fatigue syndrome, inflammatory bowel disease, psoriasis, rheumatoid arthritis, systemic lupus erythematosus, type 1 diabetes, and multiple sclerosis are among the most common.

Food allergies are also becoming much more common than they used to be. In most public schools, a child can't pack a peanut butter sandwich for lunch because an allergic classmate could go into anaphylactic shock if they came into contact with a peanut product. At least 8 percent of children in the United States have food allergies, many have multiple allergies and severe reactions.[5] Food allergies can be lethal. The percentage of children in the United States with a food allergy rose 50 percent from 1998 to 2010, according to the CDC.[6] Children are increasingly becoming allergic to the most common foods: eggs, dairy, wheat, soy, peanuts, and tree nuts.

There are several working hypotheses about what causes our immune systems to turn against us. The most popular one is that we humans have become "too clean." We talked about this in chapter 3. The idea behind the hygiene hypothesis is that public health measures, including antibiotics, pasteurization, vaccination, and water sanitation, which have so successfully limited the spread of infectious diseases, have also unexpectedly led to chronic illnesses because our extremely clean environment fails to "educate" the immune system to attack infections. Without infectious diseases or parasites to fight against, the body fights itself. The evidence to support this hypothesis is mostly epidemiologic. For example, large population studies show that allergies and asthma are much less common, even nonexistent, in countries where hygiene is poor.[7]

A study published in the *New England Journal of Medicine* in 2016 compared the risk of asthma among groups of children in two farming communities, Amish and Hutterites.[8] The Amish families practiced "traditional farming practices" whereas the Hutterites had adopted "industrialized farming practices." The Hutterite children had four times as much asthma and six times as many allergies. The researchers pointed out that the Amish children enjoyed a much more intimate relationship with animals and dirt. That is one explanation. Might it also be that the Hutterite children were exposed to more toxicants (toxins and toxic chemicals), through the industrialized farming practices of their community?

It is becoming increasingly clear that overexposure to toxicants is playing a role in the autoimmune epidemic. For example, exposure to toxins produced by mold can make people sick, partly through development of autoimmune disease. A study conducted in Finland found a greatly elevated prevalence of autoimmune disease in teachers and students who attended a moisture-damaged school. This was attributed to chronic exposure to toxic mold.[9]

I propose that glyphosate, too, is playing a significant role in autoimmune disease. My hypothesis is that, when glyphosate substitutes for glycine, it leads to a defective protein that may not even be released from the cell that synthesized it. When the protein misfolds, the body's immune cells mistake it as a foreign protein and develop antibodies to attack even healthy versions of the protein. Food proteins can also have glyphosate embedded in them, which would make it harder for them to break down in the gut. Finally, glyphosate also suppresses the innate immune system, which would ordinarily be able to clear dead human cells and pathogens that have invaded without involving antibodies. This may be the most important factor.

Immune System 101

The human immune system is a complex and miraculous biological achievement. Immune cells emerge from stem cells in the bone marrow and mature into various cell types that perform different specialized functions. Collectively our immune system is made up of white blood cells, or leukocytes, so called because they are colorless cells with no hemoglobin.

White blood cells can be broadly categorized as innate immune cells and adaptive immune cells. Neutrophils, dendritic cells, mast cells, monocytes, and macrophages are all phagocytes (literally "eating," or "engulfing" cells) comprising the innate immune system. T cells, B cells, and natural killer cells are all lymphocytes. They make up the adaptive immune system. The job of all white blood cells, collectively, is to police the body looking for trouble.

Trouble basically comes in one of two forms: an invasive pathogen (an infection of some sort) or a human cell that's been injured or deformed and needs to get cleared. The human cell could be a cancer cell. Some immune cells also have a specialized role clearing damaged materials, mostly fat and cholesterol that has been oxidized, glycated, or misfolded, or oxidized particles that can be taken up by immune cells and then broken down again into raw materials to build healthy new molecules. This includes amyloid beta plaque, which is associated with Alzheimer's disease.

Antibodies, the part of the adaptive immune system most people are familiar with, are proteins that bind to viruses or bacteria to prevent them from attacking our cells. But the innate immune system is powerful, as well. It can often clear infectious agents without ever invoking antibody production. The innate immune cells can trap pathogens and release cytokines that can kill invasive microbes. After that, our macrophages ("big eaters") essentially eat, or phagocytose, in scientific terms, the dead microbes. When the innate system can't clear invaders as fast as they are proliferating, the adaptive system steps in to lend a helping hand.

T "helper" cells are a kind of lymphocyte, which is a white blood cell made in the bone marrow that accounts for about 20 percent of the body's leukocytes. T helper cells are divided into two subgroups: Th1 cells and Th2 cells. Both Th1 and Th2 cells can release cytokines, which can be themselves divided into two broad groups: those that induce an inflammatory response (from Th1) and those that induce an allergic response (from Th2). Proinflammatory cytokines released from Th1 cells can kill bacteria, but they can also cause local tissue damage. Th2 cells produce immunoglobulins (also known as antibodies), particularly IgG, which are associated with allergies.[10]

The adaptive immune system has a sophisticated mechanism to create and then refine antibodies in response to specific antigens presented to them by members of the innate immune system. These antigens are usually proteins that are uniquely produced by an invading pathogen. When the immune system is working correctly, a population of adaptive "memory" B cells will recognize a signature protein of the invader and neutralize it. The next time that pathogen shows up, these marvelous B cells will remember and quickly mark the pathogenic cells with the antibody, like tagging a tree in an overgrown forest, so the invading cells can be easily identified by innate immune cells and removed.

The adaptive immune system is the basis for vaccines. The system creates specialized memory B cells that produce antibodies specific to the antigens in the vaccine, typically some of the proteins normally produced by the targeted virus or bacterium, or in some cases the entire pathogen in a weakened form. Interestingly, it's sometimes hard to get the body to recognize vaccine ingredients as foreign. People with healthy innate immune systems may never produce antibodies in response to vaccines.

On the other hand, people with impaired innate immunity can respond too well to vaccines. Vaccination can produce an excessive antibody response that can later lead to autoimmune diseases, through a process called molecular mimicry. This can happen when antibodies produced by the specialized B cells get confused. They attack human tissue because there is a peptide sequence within a human protein that closely resembles a peptide sequence in the antigen associated with the disease.

An important ancient component of the innate immune system is the complement cascade. This is a part of the innate immune system that helps antibodies and phagocytes clear both invaders and damaged cells from the body. It can even do the job all by itself if it is sufficiently strong. In response to an invasive pathogen or a cell under stress that's about to die, a complex sequence of events immediately takes place. This sequence involves proteins, C1 through C9, that are released from the liver and circulate in the blood in an inactive form until they're needed.

A pathogen or other trigger will cause enzymes to activate these proteins, making them capable of altering complement molecules in an elegant cascade. When C1 is activated, it launches the cascade through a series of cleavages and further protein activations in sequence, which ultimately results in several events that promote healing.[11] One of these events is opsonization, which takes its name from the ancient Greek word *opson*, meaning the side dish of a meal. Opsonization involves an opsonin, which could be a complement or an antibody, binding to an antigen to ready it for phagocytosis. Macrophages, summoned by signaling molecules, "eat" the invaders after solubilizing their membranes. They then recycle the debris.

Cells die for a number of reasons: injury, proliferation of a virus within their walls, oxidative stress, exposure to toxicants, or deficiency in a critical nutrient needed for survival. If dying cells and their debris are not efficiently cleared by the immune system, autoimmune disease can develop. This is because dead cells can release allergenic material, such as their own DNA, that then causes adaptive immune cells to produce autoantibodies that then attack other cells.

Vulnerable Peptides

In chapter 5, I talked about how glyphosate likely disrupts the enzyme NADPH oxidase in immune cells, which prevents them from releasing superoxide to kill invading pathogens. But there are many other proteins as well that the innate immune system relies on that may be disrupted by glyphosate through glycine substitution.

An important component of the innate immune system is a collection of small peptides called cationic antimicrobial peptides. These peptides, known as AMPs, are released by immune cells in response to a pathogen invasion. They work by binding to the highly negatively charged phosphate anion attached to phospholipids that are often present in the outermost leaflets of bacterial membranes, but not in human cells. AMPs need to bind to phosphate, which makes these peptides vulnerable to glyphosate substitution.

There are many different variants of AMPs. Only one amino acid in the peptide sequence has been found to be consistent among them. This is the glycine positioned in the middle of the peptide. It serves to maintain a hinge or kink between two short structures, which is essential for their function.[12] AMPs are usually positively charged, with arginine (R) and lysine (K) over-represented.[13] (Please consult table A.1, on page 186, for these one-letter codes for amino acids). Besides the glycine residue at the hinge, these peptides also tend to be rich in glycine throughout. As we know, glycine plus positively charged neighbors in a protein that binds phosphate is a recipe for disaster. Glyphosate substitution for the middle glycine residue interferes with the peptides' ability to bind phosphate, and therefore with their ability to kill pathogens.

Our Collagen, Ourselves

Collagen is considered the body's scaffolding. It is a protein that provides structure to the bones, ligaments, skin, and tendons. In China, people sometimes view it as a fountain of youth, and so they eat foods high in collagen, such as pigs' feet and shark fins, in their search for fewer wrinkles, less joint pain, and longevity. Collagen has a characteristic GxyGxyGxy . . . motif, which we talked about in chapter 4. A long sequence of this pattern folds into an elegant triple helix.[14] Because every third amino acid is a glycine, collagen is tremendously vulnerable to glyphosate substitution, which would disturb the triple helix formation.

The collagen motif is not only present in collagen; two other crucial proteins, both synthesized predominantly in the liver and released into the bloodstream, are important for complement activation, which is the cascade of biochemical events that stimulates an immune response. These are complement 1q (C1q) and mannan-binding protein (MBP). They both contain a specialized "domain," or collagen-like area that forms a stalk, which is essential for binding to the extracellular matrix of macrophages.

Six C1q molecules assemble into a triple helix formation. They then collaborate to form a hexamer that comprises the stalk. Since each C1q molecule has 26 glycines in its Gxy . . . sequence, and six molecules are assembled into each stalk, glyphosate has 156 opportunities to substitute for glycine, disrupting the protein's ability to bind to macrophages. This has serious consequences. Such a disruption impairs the ability of macrophages to clear pathogens and to remove cellular debris from dead and dying cells.

C1q proteins have another domain called the globular head that binds to a wide range of targets, including gram-negative bacteria, viruses, dying cells, products of tissue damage, immune antibody-antigen complexes, and amyloid proteins.[15] Macrophages basically use C1q as a fish hook to snag invasive pathogens and other materials targeted for clearance. C1q's collagen-like stalk binds specifically to heparan sulfate in the extracellular matrix of the macrophage, but only after it has snagged its "fish." So C1q first hooks onto the virus and then hooks the virus into the membrane of the macrophage, trapping it there. The macrophage then breaks off the piece of heparan sulfate bound to the virus and "eats" it by transporting it into its lysosomes, the cell's digestive system. Imagine a fly flying around in the room compared to one stuck on fly paper. The fly stuck to the paper is a much easier target to kill!

Mannan-binding protein binds mainly to mannan, an unusual type of sugar produced by viruses, bacteria, yeast, and fungi, but not by human cells. This makes mannan-producing pathogens a convenient target for the immune system. MBP binds to the invaders and then macrophages grab and clear them from the body. If MBP is working correctly, specific antibodies are not needed to clear the infection. But MBP, like C1q, will have trouble hooking into a macrophage extracellular matrix if its collagen-like stalk is disrupted by glyphosate. People with a genetic defect in MBP are more susceptible to acute respiratory diseases, especially in early childhood.

There's another group of proteins in the same class as C1q and mannan-binding protein that also have a collagen-like domain. These macrophage scavenger receptor class A (SRA) proteins play an important role cleaning up debris and clearing both gram-negative and gram-positive bacteria. They are promiscuous, binding to many different molecules as long as those molecules are negatively charged. Macrophage scavenger receptor proteins can bind to oxidized, acetylated, and glycated LDL, phospholipids, sulfatide, sulfated polysaccharides, lipopolysaccharides, and lipoteichoic acid. These scavenger receptor proteins act like a vacuum cleaner to suck away and clear damaged goods and pathogens.[16]

"Houston, we have a problem." When we disrupt macrophage scavenger receptor proteins, we disrupt one of the most important aspects of the immune system. Macrophage scavenger receptor proteins are able to clear bacteria and debris without invoking an inflammatory response.[17] It is the inflammatory response that eventually leads to autoimmune disease. When the innate immune system cannot efficiently trap and clear pathogens, the adaptive system is called into play. An early response is to produce antibodies to tag the invaders. Through molecular mimicry, if the invader protein resembles a human protein, the immune cells can get confused and attack the self.

A Marker of Inflammatory Disease

Some proteins' main job is to flag a problem and launch a repair mechanism. Serum c-reactive protein (CRP) is one of these flagging proteins. It is a common biological marker of systemic inflammatory disease. Doctors test the blood for CRP to find out about hidden inflammation. A healthy person will have low levels of c-reactive protein in their blood. When CRP tests come back at high levels, it can be an indication of a serious infection, including fungal infections, intestinal diseases, osteomyelitis, or sepsis. The liver responds to trauma, inflammation, and infection by producing more CRP, a lot more CRP. It can increase by as much as a factor of a thousand.[18] Glyphosate exposure induces an inflammatory response in the liver. In response, the liver will assemble many CRP proteins. This has been verified experimentally in rats exposed to Roundup, where increase in liver CRP production was dose related.[19]

This increased CRP production creates a vicious circle because the glycines in the CRP protein are vulnerable to glyphosate substitution when

CRP is produced. CRP has a sequence near the carboxyterminal end of the protein (IILGQEQDSFGGNF) that's found in many species, including mammals, fish, and even horseshoe crabs, which means this sequence is an essential part of the protein.[20] The first and last G are conserved among mammals, and the first G of the GG pair is only occasionally substituted by A (alanine), for example, in goats and cows. Even the horseshoe crab has all three Gs.

CRP is a member of a class of proteins called pentraxins, so named because they form a pentamer by assembling five copies of the protein together. Pentraxins recognize and tag a variety of pathogenic agents as well as cells in distress, allowing them to be easily eliminated through the assistance of complement pathways and macrophages.[21] CRP acts in many ways like IgG, except that, unlike IgG, it does not lead to allergies.

A characteristic feature of CRP is that it binds choline phosphate. Of course, phosphate binding is a common feature of proteins vulnerable to glyphosate substitution. The second G of the GG pair may be especially susceptible to glyphosate substitution due to its tiny left neighbor. Many bacteria that cause respiratory diseases and ear infections, such as *Streptococcus pneumoniae*, *Haemophilus influenzae*, and *Pseudomonas aeruginosa*, attach choline phosphate to their exterior coat. CRP binding to these infectious agents can help the innate immune system recognize and clear them. Defective CRP binding may result in chest infections, respiratory diseases, and ear infections.[22]

CRP also assists in the clearance of dead human cells. CRP binding to the choline phosphates on the surface of these cells activates the complement system to help macrophages clear those cells. These choline phosphate molecules are not visible on the membranes of healthy human cells. Binding of CRP to cells that are shutting down allows immune cells to identify them much more quickly and clear them before their contents spill out, which can lead to autoimmune disease over time.

In 2000, researchers at Cornell University's medical school summarized the effects of CRP: "The anti-inflammatory effects of CRP required C1q and factor H and were not effective once cells had become necrotic."[23] (Factor H is a complex regulator of the complement system.) In other words, both C1q and CRP are required to clear sick cells effectively. Necrosis, or cell decay, sets in when the cell is left uncleared, due to defective CRP and/or defective C1q. Both of these important immune mediators can be disrupted by glyphosate.

CRP and its helpers act in concert to promote clearance of defective cells in a noninflammatory way. According to the researchers at Cornell this "may help to explain how deficiencies of the classical pathway and certain pentraxins [collagen-containing proteins involved in acute immunological responses] lead to impaired handling of apoptotic cells and increased necrosis with the likelihood of immune response to self."[24] In other words, in order to prevent inflammation, you need properly functioning versions of both CRP and C1q of the "classical pathway." When these proteins aren't working, necrosis sets in, which is what leads to autoimmune disease because of the antibodies that now must be called into play. "Immune response to self" is autoimmune disease.

Friendly Fire

A long and ugly list of chronic diseases are considered to be predominantly autoimmune diseases, and many other diseases and disorders, such as autism, have an autoimmune component. I've shown you how glyphosate exposure weakens the innate immune system. It also plays an unfortunate and pernicious role in worsening certain autoimmune disorders.

Systemic lupus erythematosus is a chronic autoimmune disease. Symptoms of lupus include a characteristic "butterfly rash," arthritis, kidney damage, skin rashes, and neurological disorders. Lupus has systemic effects on the body's organs. The immune cells produce a variety of different autoantibodies that attack the tissues. In its later stages, a condition called neuropsychiatric lupus, categorized by cognitive impairment and emotional imbalance, appears.[25] But despite extensive research, scientists haven't been able to explain the cause of this debilitating disease.

Doctors originally believed that abnormalities in the lymphocytes—the natural killer cells, B cells, and T cells that mainly form the adaptive arm of the immune system—were the primary cause of autoimmunity in lupus. Under healthy conditions, T cells and B cells will phagocytose viruses and bacteria that present foreign proteins attached to their cell membranes, even in the absence of antibodies. Upon activation, these B cells and T cells also release toxic cytokines, in an attempt to kill foreign invaders. More recently, however, blame has shifted toward the monocytes and macrophages, an essential arm of the innate immune system.[26] One of the functions of monocytes and macrophages is to present foreign, unrecognized proteins to the T cells and B cells to induce antibody production.

A core feature of lupus is that the innate immune system is unable to effectively kill and consume viruses, bacteria, and food proteins that have inappropriately entered the lymph system and the general circulation. These innate immune cells are likely impaired as a direct consequence of disrupted collagen-like stalks in the three crucial proteins I described earlier—MBP, C1q, and scavenger receptor class A proteins—as well as an inability to produce functioning antimicrobial peptides. These problems are compounded by defective CRP contaminated with glyphosate. The innate immune cells recognize that there is a problem and they can't fix it. They raise alarm bells to induce the B cells and T cells to produce antibodies to tag the invaders, in order to get the invaders noticed and killed by the clean-up crew.

Neuropsychiatric symptoms affect nearly half of the people with lupus. Symptoms are diverse, but the most common are headaches, depression, anxiety, and cognitive dysfunction (what most of us think of as brain fog). The primary causes of these symptoms include antibody-mediated neurotoxicity, vasculopathy due to antiphospholipid antibodies, and cytokine-induced neurotoxicity.[27] These symptoms can all be expected when the innate immune system is sleeping on the job and the adaptive immunity system takes over in fighting infections and clearing dead cells.

Neuropsychiatric lupus can't happen unless the blood-brain barrier starts leaking. Lupus antibodies that invade the cerebrospinal fluid are able to enhance the response of receptors in the brain and cause excessive neural stimulation and neuroexcitotoxicity, which is likely the underlying cause of the neurological symptoms. This implicates glyphosate in two more ways: Glyphosate induces leaky barriers; and glyphosate has been shown in multiple studies to excite NMDA receptors, both by acting as a glycine analogue and by suppressing the clearance of glutamate from the brain's synapses.[28] With excessive NMDA excitation, calcium builds up internally and eventually causes catastrophic failure of the mitochondria.

Lupus is also associated with an increased risk of cardiovascular disease. Inflammation of the artery wall damages the endothelial cells lining the blood vessels. When these cells die, they release blebs (blisters, basically) at their surface, which would ordinarily bind to C1q in order to attract macrophages to come in and clear the debris. But with a defective stalk, C1q is unable to bind to macrophages. The exposed debris accumulates and induces an autoimmune response.[29]

Unfortunately, that's not all. Lupus is also associated with skin problems, particularly extreme sensitivity to sun exposure. People suffering from lupus who spend time in the sun will get red spots on their skin and a scaly, purple rash on their faces, necks, and arms. This uncomfortable and embarrassing reaction is likely caused by lupus antibodies to a molecule called calreticulin.[30] Calreticulin is normally retained inside the cell unless the cell is injured. Upon exposure to UV light, skin cells, called keratinocytes, can become injured. They then release their calreticulin into their membranes, presumably as a signal to macrophages to orchestrate the clean-up process. In healthy people, keratinocytes are constantly dying and getting replaced.

Normally, C1q's globular head binds to calreticulin when calreticulin is exposed on the surface of cells. This is how macrophages initiate the complement cascade. But if C1q is defective because it has glyphosate embedded in the stalk, for example, it will be impaired in its ability to clear calreticulin-tagged dying cells. I hypothesize that in people suffering from lupus, the normal process to clear dying keratinocytes is defective. Unlike the healthy C1q-based mechanism, the antibody-based mechanism is associated with an inflammatory response that causes the red spots and scaly skin.

The experience of a young man with severe C1q deficiency provides insight into how glyphosate damages proteins. This patient was suffering from a genetic form of lupus. He had a glycine mutation at residue 34 in the stalk region of C1q, inherited from both his parents. Because of his genetic condition, arginine substituted in for glycine. This one change resulted in a defective version of the protein that was never even released from the liver cells that synthesized it, leading to a severe pathology involving lupus. The poor young man, at age 24, had skin issues, weakness in his arm, impaired vision, intermittent fevers, canker sores, sunlight sensitivity, memory and concentration problems, and a seizure disorder.[31] All because of a single glycine molecule that had been replaced with something else. We can predict a similar outcome, albeit to a lesser degree, when glyphosate substitutes for this critical glycine residue in C1q.

Celiac Disease

Gut issues of all kinds, including among young people, have been increasing, though mainstream doctors have been unable to figure out why. Celiac disease, colitis, and other inflammatory bowel conditions are becoming global problems.[32]

Celiac disease is a digestive disorder that causes damage to the villi in the small intestine. It's caused by autoantibodies that develop against peptides derived from gluten, which is found in barley, rye, and wheat. Although there is no Roundup Ready wheat on the market, wheat is often highly contaminated with glyphosate because farmers use it as a desiccant on wheat crops shortly before harvest. There is no question that celiac disease is far more prevalent today. Samples of blood serum collected between 1948 and 1954, compared to samples collected today, reveal at least a *fourfold* increase in antibodies against gluten since then.[33] An even larger number of people suffer from gluten intolerance without an explicit celiac diagnosis.

Gluten intolerance results when multiple small peptides derived from gluten remain undigested. Many of these peptides terminate in proline, suggesting that the enzyme that detaches proline is defective. Currently available digestive enzyme supplements are ineffective in degrading immunogenic gluten epitopes, the part of the protein that an antibody would attach to, due to blockage at proline residues that prevents further disassembly of the peptide chain.[34]

Gliadin is a major protein component of gluten, and the primary target of gluten antibodies in celiac disease.[35] It has a high concentration of both glutamine and proline, and both play a role in the disease process. Proline is the only coding amino acid that has a side chain on the nitrogen atom. This makes it difficult to break it apart from its neighbors in the peptide chain. As a consequence, specialized enzymes have evolved that deal specifically with proline.[36] *Lactobacillus* bacteria have multiple enzymes that specialize in detaching proline from peptide sequences. These hydrolytic enzymes, which use water to break a bond, aren't present in human cells. They have a motif surrounding the serine residue at the active site that contains three glycines: GxSxGG (Glycine-x-Serine-x-Glycine-Glycine).[37] We can expect all these hydrolytic enzymes to be vulnerable to disruption by glyphosate at one or more of these glycine residues. *Lactobacilli* are also sensitive to glyphosate, so their numbers in the gut are jeopardized by glyphosate exposure.

In 2019, an international team of 21 scientists found that gluten intolerance is associated with an increased expression of a microbial protease (an enzyme that breaks down proteins) called elastase that can metabolize gluten. They also found that gluten intolerance is associated with an overabundance of the pathogenic species *Pseudomonas aeruginosa*, a rod-shaped bacterium that expresses elastase.[38] Impaired gluten metabolism likely allows undigested gluten proteins to accumulate in the colon, which then supports the overgrowth of species producing elastase. Using a mouse model, these authors showed that gluten exposure led to *Pseudomonas* overgrowth, and induced an inflammatory response leading to villus blunting, or flattening of the surface. Microbial metabolism of undigested proteins in the colon causes increased ammonia production along with a rise in pH, disrupting acetate synthesis. (As you may remember, acetate is a precursor to acetyl coenzyme A, a molecule that feeds into the citric acid cycle to generate energy for the cell.) *Pseudomonas aeruginosa* is one of the few bacteria that can fully metabolize glyphosate, giving the bacteria a competitive advantage under conditions of chronic glyphosate exposure. Infection with multiple-antibiotic-resistant *Pseudomonas aeruginosa* has become a problem in hospitals in the United States.[39]

In addition to disruption of gliadin metabolism due to glyphosate contamination in enzymes that process proline, another factor contributing to celiac disease is a deficiency in the clearance of apoptotic cells. Defective scavenger receptor expression is linked to celiac disease, as are defects in mannan-binding protein (MBP).[40] Inefficient clearance of apoptotic cells likely leads to immune activation that then causes damage to the intestinal villi. Glyphosate substitution in the collagen-like stalk region of MBP can cause defects in both the MBP and the scavenger receptor.

Glyphosate and COVID-19

COVID-19 is a novel infectious viral disease that became a global and devastating pandemic. However, there is a huge difference in the degree to which this virus affects people in different countries. Many of the hardest hit countries, such as the United States, the United Kingdom, Brazil, and South Africa, make heavy use of glyphosate. Diabetes, obesity, and hypertension, diseases that are prevalent in the industrialized nations, are well-established risk factors for mortality from COVID-19.[41] Glyphosate use in the United States is highly correlated with the rise in these conditions.[42] It is becoming clear that a major factor leading to severe disease is an overzealous response by the adaptive immune system, leading to increased antibody production and a cytokine storm.[43] This response can be expected because of glyphosate's severe disruption of the innate immune system, which would ordinarily play a critical role in clearing the virus. Finally, there is increasing awareness that patients who recover from COVID-19 may develop autoimmune disease consequently, due to overproduction of antibodies to viral proteins that attack the host through the mechanism of molecular mimicry.[44] Few researchers are talking about glyphosate in the context of SARS-CoV-2. But they should be.

Chronically Exhausted

Chronic fatigue syndrome is a common but poorly understood disease. People suffering from chronic fatigue experience cognitive impairments, disturbed sleep, and physical exhaustion after exercise. It's often brought on by an infection that develops into a chronic condition after the infection resolves.

Some researchers argue that chronic fatigue syndrome is best understood as an autoimmune disorder developing from gut dysbiosis and a leaky gut barrier.[45] Microbes that would normally stay in the gut migrate into the general circulation, causing an immune response. This results in sustained production of memory B cells that produce antibodies in response to the antigens produced by the microbes. On top of that, viral infections, particularly the Epstein-Barr virus, which is often associated with the disease, can lead to the development of permanent antibodies. All these antibodies run the risk of becoming autoantibodies that attack the body's own proteins that are essential for energy production in the mitochondria.

Interestingly, one person who contracts Epstein-Barr might experience no ill effects while another who contracts it might suffer chronic fatigue for years. Why might this be? Could an overexposure to environmental toxicants prompt the body to misstep into autoimmunity? Could glyphosate's disruption of the gut microbiome, induction of a leaky gut barrier, and suppressive effect on innate immunity be the trip wires to induce susceptibility to chronic fatigue and other syndromes?

We are seeing a rise in autoimmune disorders not only in humans but also in pets. Dogs suffering from autoimmune thyroid disorders are sometimes prescribed thyroid replacement hormones. Canine lupus is also becoming increasingly common, as is canine inflammatory bowel disease (IBD).[46] Cats, too, suffer from feline systemic lupus erythematosus, which can have a variety of symptoms, including skin lesions and lameness. Feline autoimmune disorders, including arthritis, IBD, and immune-mediated hemolytic anemia, are also on the rise. Even horses are experiencing autoimmune diseases.[47]

As I mentioned previously, there is no question that environmental toxins and toxic chemicals are playing a role in provoking autoimmune diseases. It's not just that we may be "too clean," as the hygiene hypothesis suggests. It's also that we are overmedicated and overexposed to heavy metals and other cell-damaging toxins, such as those from mold. Both medical and animal doctors have pointed to the overuse of antibiotics and over-vaccination as probable triggers for autoimmune disorders. But glyphosate, through its disruption of the gut microbiome and its insertion in place of glycine in diverse proteins, impairing their folding and their function, is also a major culprit.

Once you understand the mechanisms by which glyphosate is causing damage to human (and microbial) health, you see how the presence of glyphosate in living tissues can exacerbate health conditions that may have initially been triggered by something else. Autoimmune diseases are like a fire that the body is struggling to get under control. Sometimes they seem to go out—you can go for weeks or even months or years without symptoms—and then they flare up again, causing debilitating problems and pain. What stokes the fire of these relapsing and remitting diseases? The conventional answer is "we don't know." But we do. We know that the more often life on Earth is exposed to glyphosate, the worse it is for all of us.

— CHAPTER 11 —

Reboot Today for a Healthy Tomorrow

Live in the sunshine. Swim the sea. Drink the wild air.

—Ralph Waldo Emerson

I receive emails and letters from people around the world, every day. They write to share their stories with me. Some have so many food sensitivities and allergies that they struggle to find anything they can eat. Others describe strange and debilitating health conditions that have flummoxed their doctors for years. Parents of children with autism and other neurological disorders share their despondency and ask for my advice.

I am grateful for these letters. I'm heartened by the dogged determination people have to understand what's making them sick; by their eagerness to dive into complex science they never learned in school; and by the humility and openness it takes to change eating and lifestyle habits they may have had for decades. You're doing better than you think. Even your willingness to ask questions, and to read a book like this, is a step in the right direction. We are up against Goliath. Because the truth is that there are powerful forces at play that profit, enormously, from making people, including children, very sick.

Glyphosate is a global threat, and we cannot be satisfied until it is banned worldwide. The agrichemical industry's persistent deniability of harm is nothing short of criminal. Manufacturing plants where glyphosate is being synthesized need to be shut down. Research dollars need to be spent on projects that investigate how to get glyphosate out of drinking water, out of rivers and oceans, out of soil, and out of our bodies. Once glyphosate is gone for good and its residues are no longer detected in our soil, food, water supply, and urine, we will know that we have won. It's time to change. We need solutions for producing food using renewable agriculture, practices that improve the soil year by year instead of contaminating it. We must never make the mistake we made with DDT and simply substitute glyphosate for other toxic herbicides.

I believe that glyphosate is the most dangerous environmental chemical we face today due to its unique mechanism of toxicity, careless application, and pervasive presence. But other toxicants are destroying our health and poisoning our planet, as well. These include aluminum, arsenic, bisphenols like BPA, fluoride, lead, mercury, phthalates, polyvinyl chloride, polyurethane, Teflon, styrene, and many others.[1] And of course there are other toxic herbicides, insecticides, and fungicides that are commonly used in agriculture. As I've explained throughout this book, many of these chemicals work synergistically in a way that amplifies their toxicity. Sometimes the burden to the body is so great it results in infertility. Even when it doesn't, later generations can be harmed by exposures their parents or even their grandparents experienced when they themselves were young.

If you're like me, it's hard to fathom that entire industries, or even whole sectors of the economy, could be predicated on human sickness. It's simply hard to believe. Perhaps the most dissonant realization we must confront is that the health care sector is predicated on making and keeping people sick. When you follow the money, you realize that the pharmaceutical industry, and most of the medical establishment, make the bulk of their profits from treating symptoms of chronic illness. The pharmaceutical industry has experienced tremendous growth in the past 20 years; the research, production, and distribution of medication is so lucrative that it generates $1.25 trillion in revenue in a single year.[2] Autoimmunity alone is a $108 billion industry.[3] There is little incentive to identify and correct the root causes of chronic disease or empower people to keep from getting sick in the first place when there's so much profit to be made. Quite the opposite, in fact. The pharmaceutical industry thrives when America is unhealthy. Vibrant good health harms its bottom line.

That's the bad news. And it's easy to get lost in the weeds of despair thinking about how unchecked human activity is poisoning plants, animals, waterways, and ourselves. But there's also a lot of good news. One by one, people are waking up and realizing the urgency of the crisis we face, recognizing that we must focus on improving the soil and the nutritional value of food rather than just on yield. Small organic farmers are the superheroes of the twenty-first century, as are the health advocates, political activists, and lawyers who are fighting Bayer AG and other chemical companies and urging lawmakers to stop turning a blind eye.

Regenerative techniques can heal the Earth, arresting and even mitigating climate change by sequestering carbon in the soil. We can identify organic products that reduce weed growth without poisoning the Earth or the organisms that live in and on the soil. The human immune system functions best when the human it is protecting is well fed and toxicant-free. If food crops are well supplied with nutrients, they, too, are hardier, better able to withstand fungus growth and insect attacks. This creates a cycle of health, reducing the need for fungicides, insecticides, and herbicides. We have the individual and the collective responsibility, as well as the power, to stop the worldwide use of glyphosate.

As the "sick care" sector serves its own ends to keep people chronically ill, many alternative medicine practitioners are finding new ways, or rediscovering ancient ways, to help people heal using foods, herbs, and other natural products instead of synthetic drugs that disrupt biological pathways. I envision a future, hopefully soon, when humans become shepherds of the Earth and people enjoy strong and healthy lives free of pain and full of joy. We have the capacity to pull this off. We just need the will.

My goal is to open your eyes to the science that implicates glyphosate. As overwhelming as some of this is, this information is a call to action, not a reason to despair. When we know better we do better. We don't have to leave a toxic legacy to our children and our children's children.

The biochemistry of our bodies is marvelous and miraculous. Much of the damage caused by glyphosate and other chemicals is reversible. Think of your health as a barge that's been slowly heading out to sea in the wrong direction. It takes a lot to turn it around. It won't happen overnight, but it will happen. And once you're back on track with your health firmly in your own hands, you'll be amazed at how much more energy you have, how much clearer your skin is, how much sharper your brain becomes, and how much better you feel. Knowledge is power. There are simple, powerful changes you can make right now that will make a world of difference.

Eat Real Food

The most important step you can take toward a long and healthy life is to eat wholesome, nutritious, real food. If you're fortunate enough to have access to farmers markets, community supported agriculture (CSA) networks, food cooperatives, or your own garden, that's your best bet for knowing what your food is, how it was grown, and if anything poisonous was applied to it. If you don't have access to this level of real food, and many people don't, look for the "Certified Organic" label (if you're in the United States) wherever you buy your food. Certified Organic labeling is not perfect, and it's no substitute for buying real food from a local farmer, but it is a simple way for consumers to choose foods that are less likely to be contaminated with glyphosate. Also, genetic modification is not allowed in certified organic foods. Wherever you buy your food, do your best to have the bulk of your diet be whole real food instead of packaged food-like substances. Even when packaged products are labeled as organic and GMO-free, they still aren't necessarily healthy foods. Broken down into basic nutrients like sugar, flour, and oil, packaged organic food leaves out all the complex organic molecules and many of the micronutrients that offer many benefits to health, some of which are not yet even understood.

Journalist and food writer Michael Pollan recommends frequenting the perimeter of the grocery store. Stock up on fresh vegetables, fresh fruits, wild fish, and grass-fed, grass-finished organic meats, if they're available. Avoid boxed foods with long lists of ingredients, especially if the ingredients are unrecognizable to you; packaged foods often include mold inhibitors, artificial dyes, and artificial colorants. And if you feel overwhelmed, start small and see what's available. Grab a few organic avocadoes one week to make guacamole. Buy a quart of organic plain whole milk yogurt and make some fruit-sweetened smoothies. Get the good eggs, from chickens with access to lots of outdoor space, instead of the cheap eggs. They *are* more expensive than nutrient-deficient conventional eggs but they are still cheaper than a candy bar. The gift of lifelong good health is a marathon, not a sprint.

While there is a confusing array of dietary fads claiming that low-fat, or high-fat, or low-carb, or high-protein is the best, what matters more is the micronutrient density in your diet. We need not only vitamins and minerals but also the polyphenols and flavonoids in fruits and vegetables, as well as in herbs and spices. Many of the herbs that we think of as flavoring are actually foods that promote health and longevity. These include oregano, rosemary, basil, cilantro, dill, sage, parsley, turmeric, garlic, and ginger. These foods add extra flavor and are also vitally important for supporting metabolism and detoxification. Consider this: In a study of over 11,000 people, scientists discovered that people who have the healthiest guts eat more than 30 different types of plants every week.[4] Your mom was right when she told you to eat your vegetables (as long as they're not contaminated with glyphosate and other toxicants). What we all need is a rich variety of different foods across the macrobiotic spectrum of fats, carbohydrates, and proteins, naturally flavored with abundant organic herbs and spices.

Savor Sulfur

Since a deficiency in dietary sulfur, and especially in sulfate, is a factor in many modern diseases, it's important to enjoy lots of sulfur-containing foods. Animal-based protein has higher levels of sulfur-containing amino acids—taurine, methionine, and cysteine—than plant-based proteins. Seafood, grass-fed beef, fish, eggs, and cheese are all good sources of sulfur. Among plants, onions, leeks, garlic, and cruciferous vegetables such as cabbage, Brussels sprouts, broccoli, and cauliflower are the best sources of sulfur-containing organic molecules. Garlic, which is a food you should be eating often, contains allicin, an organosulfur compound that gives garlic its unique odor and is associated with many health benefits. While people tolerate these foods differently, all of them are packed with nutrients and fiber and are extremely healthy.

Cruciferous vegetables contain sulforaphane, an unusual organic compound that contains two sulfur atoms. Broccoli sprouts contain a high concentration of sulforaphane. Sulforaphane is produced only in the seed, so a single broccoli sprout contains as much sulforaphane as a full-grown plant. In 2009, a team of researchers in Italy found that sulforaphane leads to an increase in glutathione in neurons, protects them from oxidative stress, and reduces indicators of apoptosis.[5] Parkinson's disease is associated with the loss of dopaminergic neurons. Sulforaphane seems to help safeguard these neurons. Indeed, a more recent study on humans suffering from schizophrenia confirmed that people with schizophrenia have low levels of glutathione in the brain. This study also found that supplements with sulforaphane increased brain glutathione.[6]

Drinking water can be a good source of sulfur. Places where there is a lot of basalt rock, such as the volcanic islands of Iceland and Japan, are naturally blessed with high levels of sulfur that feeds into both the food crops and drinking water. Unfortunately, water processing such as reverse osmosis to remove glyphosate also removes sulfur.

N-acetyl cysteine (NAC) is a sulfur-containing amino acid. A study on rats exposed to glyphosate-based herbicides that were also given NAC showed that it had a prophylactic effect. Supplementation of NAC protected glyphosate-exposed animals from decreases in glutathione levels in the blood, liver, kidney, and brain normally induced by glyphosate. Glyphosate also induced increases in malondialdehyde, which is an indicator of oxidative stress. Simultaneous supplementation with NAC improved this metric, as well.[7]

Other sulfur-containing supplements include α-lipoic acid, chondroitin sulfate, glucosamine sulfate, Epsom salts, methyl sulfonyl methane (MSM), S-adenosylmethionine (SAMe), liposomal glutathione, taurine, and garlic. If you soak in a tub of Epsom salt (magnesium sulfate), you absorb sulfate through your skin. Even better is to bathe in natural hot sulfur springs. Absorbing sulfate through the skin bypasses the complexities of sulfur metabolism in the gut, which can be especially problematic in the presence of chronic glyphosate exposure. Glyphosate causes an overgrowth of sulfur-reducing bacteria in the gut that can produce excessive amounts of hydrogen sulfide gas, causing bloating and abdominal pain, as well as brain fog.

In order to properly use sulfate in the body, it is not enough just to be able to synthesize it. Free sulfate levels in the blood have to be maintained at a low concentration in order to prevent gelling of the blood. Many herbs and spices have great nutritional and medicinal value in part because they facilitate sulfate transport in the blood and sulfate delivery to the tissues. These include berberine, cinnamon, curcumin, resveratrol, and other flavonoids and polyphenols.

Heal the Gut

Prebiotics are nondigestible plant fibers. These complex carbohydrates, which cannot be digested by the human digestive system, are fermented by the beneficial bacteria that live in our intestines (also known as colonic flora). Prebiotics, such as oligosaccharides and lactulose, can reduce the pH of the colon and increase butyrate and calbindin, which increases the absorption of minerals and trace elements. These beneficial changes are mediated by fermentation of the prebiotics by gut microbes residing in the colon.[8] When we eat mostly processed foods, our bodies become deficient in prebiotics.

Eat foods that are high in prebiotics at *every* meal. These foods include asparagus, artichokes, bananas, dandelion greens, garlic, leeks, and onions, all of which are wholesome and delicious. Mycotoxins produced by toxic molds are present in many foods, but prebiotics can help to boost the supply of gut microbes able to metabolize these mycotoxins to less toxic metabolites.[9] Benefits of prebiotics include an increase in *Bifidobacteria* and *Lactobacilli*, a decrease in pathogenic species in the intestines, production of beneficial metabolites, improvement in calcium absorption, reduction in protein fermentation and ammonia production, improved gut barrier function, and improved immunity.[10]

Probiotics are living bacteria that promote human health. They can be found in some naturally fermented raw foods such as kimchee, sauerkraut, plain yogurt, and kombucha. They can also be found in supplements, though which brands containing which species of beneficial bacteria that are helpful is the subject of ongoing scientific debate. *Bacillus* species are a common choice for probiotics because they thrive in the human digestive tract and they are able to survive the acid of the stomach through spore formation. The spores germinate in the small intestine and can then colonize the gut. A study involving an eight-week dose of *Bacillus subtilis* probiotics showed that inflammatory markers TNF-α and IL-6 were markedly reduced following treatment.[11]

Probiotics, which are increasingly recommended to patients even by mainstream health practitioners, have come under intense scientific scrutiny. Our bodies are home to trillions of microbes. Most probiotics contain at least one of over a dozen strains of bacteria. But which live microorganisms and in what amounts should we consume for them to actually be of benefit? A study on the health benefits of a formulation containing five strains of spore-forming *Bacillus* along with a prebiotic fiber blend produced remarkable results consistent with the idea that these microbes are clearing glyphosate from the gut.[12] The experimental setup involved a simulation of the human gastrointestinal tract. Addition of the supplement resulted in increased microbial diversity in the simulated colon, with an increase in representation from *Lactobacillus* and *Bifidobacteria*, an increase in acetate levels in the distal colon, as well as increased butyrate and lactate concentrations along the entire colon. Since glyphosate reduces acetate levels in the gut, and preferentially kills *Lactobacillus* and *Bifidobacteria*, these findings are encouraging.

Supplementing with probiotics has also been found to reduce the risk of necrotizing colitis (a condition that can be fatal in preterm infants), help children suffering from persistent diarrhea, and reverse some of the negative effects of antibiotics.[13]

Harness the Power of Plants

Polyphenolic compounds are metabolites produced by plants to protect themselves against disease, infection, and damage. They can be subdivided into four primary classes: flavonoids, stilbenes, phenolic acids, and lignans. Many polyphenolic compounds have been found to help prevent diseases, including cancer, diabetes, cardiovascular disease, and neurodegeneration. They are present in a wide variety of plant-based foods, including fruits, vegetables, herbs, cereals, tea, coffee, nuts, seeds, and beer. Some that you may be familiar with are quercetin (in green tea), berberine (from the barberry bush), resveratrol (found in grape leaves and wine), and turmeric (in curry powders).[14] Polyphenols seem to protect against cancer by preventing endothelial-to-mesenchymal transformation, a crucial step in progression toward a tumor.[15] A diet consisting primarily of heavily processed foods is deficient in polyphenols.

Researchers are puzzled as to exactly how polyphenols work in the human body, mainly because they rarely make it into the general circulation. They are circulated between the gut and the liver, and the liver modifies them through conjugation with sulfate, glucuronate, and methyl groups. I suspect that polyphenols act as carriers of these biologically useful molecules. Their phenol-based rings are likely crucial for enabling efficient transfer of their conjugated unit to another molecule in the glycocalyx. In particular, they would serve to enhance the supply of sulfate and glucuronate to the mucins in the gastrointestinal wall, supporting gut health.

When avocados are in season, my husband Victor and I often make a delicious guacamole as the main dish for our lunch. We spice it up with tomatoes, coriander, onions, garlic, jalapeño peppers, and fresh lime juice. The avocado provides healthy polyunsaturated fats that are good for the brain, as well as folate, fiber, vitamin K, and copper. The other ingredients, especially the coriander, garlic, onions, and peppers, provide a wide array of complex polyphenols, terpenoids, and flavonoids. There's vitamin C in the tomatoes and lime juice, and sulfur in the garlic and onions. And so much tastier than a vitamin pill!

Eat Your Antioxidants

Glutathione and vitamin C are important antioxidants in both the liver and the blood, as well as other parts of the body. As you know, glyphosate disrupts the supply of glutathione in the liver, as does acetaminophen. Since glyphosate and many other toxicants induce oxidative stress, it is important for us to keep our bodies' supply of these two antioxidants as high as possible.

Glutathione is found naturally in some foods. Asparagus, avocados, okra, and spinach are some of the richest dietary sources. But you can also boost your glutathione levels by eating foods that are rich in sulfur-containing amino acids, such as taurine, cysteine, and methionine. Good options include nuts and seeds, as well as beef, cheese, chicken, duck, eggs, fish, pork, soy, and turkey. Assuming the enzymes needed to synthesize glutathione from these precursors are functioning properly, the body can make glutathione from these sources. Since glycine is also a component of glutathione, eating glycine-rich foods such as organic beef broth can also help.

Eating foods that are rich in vitamin C will reduce your need for glutathione, because vitamin C is also an effective antioxidant. Many fruits, particularly grapefruit, lemons, limes, and oranges, are rich in vitamin C. Vegetable sources are, too. Broccoli, Brussels sprouts, and kale, as well as herbs and spices such as parsley and thyme, are surprisingly rich in vitamin C. The takeaway message here is to use fresh spices liberally to make your food more delicious and nutrient dense. Enjoy fresh fruits as often as possible, preferably at every meal. Just make sure the fruits and vegetables you are eating are either certified organic or grown without pesticides and herbicides. You don't want the benefits of these powerful foods to be undermined.

And Don't Forget the Minerals

Glyphosate is a major metal chelator, particularly for +2 cations such as iron, magnesium, manganese, cobalt, copper, and zinc. As such, glyphosate makes these essential nutrients less available to exposed crops, which translates to deficiencies in the foods. I don't recommend specific mineral supplements because taking a lot of supplements can lead to biochemical imbalances, but products are available that offer multiple minerals packaged up into a single supplement that may help offset nutrient deficiencies in the food supply. A simple thing to do is to switch from table salt (sodium chloride) to Mediterranean sea salt or Himalayan salt in cooking. These contain multiple minerals in the balance such as they exist in nature. Eating foods that are naturally nutrient-dense and rich in minerals, such as bone broth, seafood, eggs, and organ meats, is the best option for maintaining a balanced supply of the various minerals.

Vegan Vexations

The vegan diet excludes not only all meat but also animal-derived products such as cheese, eggs, and honey. Veganism has gained considerable momentum in recent years, and it is championed not only as a healthy diet but also as a way to help save the planet. The argument is that it takes much more energy to produce animal-based foods than plant-based foods. Cows belch and expel methane (CH_4), which accounts for over 5 percent of human-generated greenhouse gases.

While these are important considerations, it's much more difficult for humans to get adequate nutrition eating only plants. A plant-based diet reduces your ability to get adequate amounts of sulfur-containing amino acids. Just as worrisome, cobalamin is completely absent in a plant-based diet. And there is no taurine in a plant-based diet unless you enjoy the edible seaweed found along the coasts of the Caribbean, Japan, and Southeast Asia. Red algae, which includes mafunori, fukurofunori, kabanori, and ogonori (or ogo), are relatively high in taurine.[16] Green seaweed has become a popular snack for children of health-minded American parents, but, unfortunately, brown and green algae contain no taurine. As much as I encourage you to eat plants, lots and lots of plants, I don't think a strictly vegan diet promotes health, well-being, or longevity in most humans.

Should Humans Be Eating Cows?

It's hard to go wrong if you prepare your own home-cooked meals using fresh unprocessed ingredients that are sourced from local organic farms. I believe this is the most important step you can take to promote health and longevity. But one of the foods that some champion and others insist is harmful is beef. Sally Fallon Morell is the founding president of the Weston A. Price Foundation, an organization that promotes traditional foods that were widespread prior to twentieth-century industrialization. Fallon Morell advocates a return to nutrient-dense diets that include raw milk, animal fats, organ meats, beef broth, and fermented foods. As a big fan of her work, I have attended the Wise Traditions Conference every year for the past 10 years. One of the foods Fallon Morell promotes is grass-fed beef. Grass-fed beef is a source of many valuable nutrients such as heme iron, taurine, B vitamins (particularly cobalamin and choline), minerals, and healthy fats.

There is an important distinction to understand. Factory-farmed cows confined to small pens are fed predominantly GMO Roundup Ready corn and soy products. These animals suffer for their entire lives. They're then herded through a chute to be weighed, put into holding pens, and taken to a kill floor. A cow that's frightened before it is butchered has so many stress hormones in its bloodstream that it can affect the taste and the color of the meat. And, as I mentioned in chapter 7, American cattle farmers frequently administer ractopamine, which is so dangerous it's been banned by more than 160 nations, to their farm animals to keep the meat lean. Factory farming is not beneficial to our health or our planet's ecosystem. It's also not humane.

Pasture-raised animals are a different story. Cows that free range over a large area make manure that naturally fertilizes the grass, and their grazing stimulates its growth. This makes the grass healthier and, in turn, increases the amount of carbon that is converted to organic matter in the soil. Humans are unable to digest grass, so any beef derived from grass-fed cows is a way for humans to indirectly convert grass into a food source. There are vast acres of grass growing in regions where farming is not possible due to insufficient rain or poor soil quality. But this acreage is perfect for grazing cows. Grazing cows have a much better life than factory-farmed cows. They belch and expel less methane gas because their diet is healthy. As opposed to manure from confined animals, which is a toxic waste product that pollutes the soil and ground water, the manure of grass-fed animals makes an excellent natural fertilizer.

The vegan equivalent of a hamburger is not a healthy or sustainable option compared to hamburgers made from natural grass-fed beef.[17] One brand of vegan burgers, the Impossible Burger, is made by growing genetically modified yeast, which has been engineered to produce a soy-based vegetarian heme product called soy leghemoglobin. This is not something humans have evolved to eat. The engineered molecule is added to soy in order to produce a vegan burger with a meaty taste. But soy monocropping using chemical-based agriculture is not good for the planet. Despite the hype, major advertising, and a state-of-the-art interactive website, eating an Impossible Burger is not superior, ecologically, to eating a grass-fed beef burger.

The Beyond Burger, manufactured by the Beyond Meat company, is another fake-meat burger with ingredients that include pea protein isolate, canola oil, coconut oil, yeast, natural flavors, gum arabic, sunflower oil, salt, succinic acid, acetic acid, non-GMO modified food starch, cellulose from bamboo, methylcellulose, potato starch, beet juice extract, citrus fruit extract, and vegetable glycerin.[18] It's best to avoid Frankenfood like this. While both the Impossible Beef burger and the Beyond Meat burger contain glyphosate, the watchdog organization Moms Across America found 11 times as much glyphosate, which is made with Roundup Ready soy, in the Impossible Beef burger.[19]

The Importance of Sunshine

Most Americans, as well as humans around the world, are increasingly deficient in vitamin D. Although we are not photosynthesizers like plants, our bodies utilize sunlight to synthesize this important fat-soluble hormone. Sunlight exposure, to both the skin and the eyes, is vital to health and well-being. We have been told for so many years without nuance that the sun is toxic. But the sun has always been a resource for planet Earth. Just as sunlight is essential for plants to grow, it plays an essential role in energizing animals, including humans.

People who live in places with little sun have a higher risk of many chronic conditions, including multiple sclerosis, diabetes, cardiovascular disease, autism, Alzheimer's disease, and age-related macular degeneration.[20]

When a team of Swedish scientists analyzed the behavior of nearly 30,000 women in Sweden, following them for almost two decades, they found that the women who lived longer and had fewer heart problems were the ones who spent *more*, not less, time in the sun. Those who actively avoided the sun had a twofold higher mortality rate compared to those who actively sought sunlight exposure.[21]

One of the most important things you can do to improve your overall health is to get out in the sunlight every day for at least a half hour, even more if you have dark skin, without sunscreen or sunglasses. Humans evolved in a world where the sun was always available, and biology never ignores a chance to capitalize on an energy source. The skin is essentially a solar-powered battery. Sunlight is translated to energy by the extraordinary physical properties of water. Infrared light grows exclusion zone water by at least fourfold.[22] UV light energizes water molecules to mobilize protons and electrons. The electrons induce the production of superoxide from oxygen. Sulfur is oxidized by superoxide to produce sulfate through the sequential action of eNOS in the membrane and sulfite oxidase. The resulting sulfate anion captures energy from the sunlight in a useful form, and this energy is vital for the health of circulating blood.[23]

Most people see sunlight as useful only for the production of vitamin D. They figure they can just take a vitamin D supplement and avoid the sun. While over a hundred peer-reviewed scientific studies have demonstrated that people with high levels of vitamin D are at lower risk of a long list of debilitating diseases, controlled studies involving vitamin D supplements have been disappointing. Vitamin D is well known for its ability to build strong bones and teeth but a three-year study published in 2019 involving patients who took three different amounts of vitamin D supplements (400 international units [IUs] per day, 4,000 IUs per day, and 10,000 IUs per day) showed that those on the highest doses of vitamin D actually had significantly *worse outcomes* in bone mineral density.[24] Vitamin D is a signaling molecule. When you get it from the sun, you also get a boost in cholesterol sulfate. If you take vitamin D in supplement form, you don't get the concomitant increase in cholesterol sulfate. So the vitamin D essentially signals with a false message, causing cells to respond inappropriately, which may negate some of the potential benefit.

Consider this: Indian researchers tried two different interventions to treat a population of a hundred men diagnosed with vitamin D deficiency at onset. Half the men were "treated" with sunlight exposure for 20 minutes midday, every day, and the other half were treated with a vitamin D_3 supplement of 1000 IU/day.[25] While those taking the supplement had a larger increase in serum vitamin D post treatment, they also had a significant increase in serum LDL cholesterol, the indicator that prompts doctors to prescribe a statin drug. On the other hand, those treated with sunlight exposure saw their cholesterol levels drop. The reduction in serum cholesterol with sunlight exposure is likely due to the increased availability of cholesterol sulfate, produced in response to sunlight exposure in the skin, facilitating cholesterol delivery to the tissues. This explains both the drop in serum cholesterol and the benefits of sunlight to cardiovascular disease prevention found in population studies of those who live in sunny, versus rainy, regions of the world.

Oh, and Ditch the Sunscreen

Malignant melanoma is a lethal form of skin cancer. Over the past 50 years, incidence of melanoma has risen faster than that of almost any other cancer.[26] Melanoma accounts for the majority of deaths from skin cancer. Over this same time period, the use of sunscreen has skyrocketed. You can't go to a beach in America, or even a park, without seeing people slathering thick layers of sunscreen and sunblock all over their skin. Sunscreens today offer much higher sun protection factor (SPF) levels: as high as SPF 64, which is 10 times the amount used in the 1960s when my children were growing up. The main justification for the widespread use of sunscreen is to protect us from skin cancer. Yet increasing usage correlates strongly with increasing rates of the deadliest form of skin cancer.

Why are skin cancer rates climbing? Could sunscreen be a causative factor? Perhaps. Zinc oxide (ZnO) nanoparticles are commonly found in sunscreens. They work by helping to block UV rays. But ZnO is cytotoxic to human cells grown in culture, even in the absence of sunlight exposure.[27] Cells exposed to concentrations ranging from 8 to 18 μg/ml of ZnO exhibit membrane leakage and oxidative stress. These cells have elevated levels of reactive oxygen species, depleted glutathione, and increased damage to cellular fats. Some scientists argue that the ingredients in the sunscreen are oxidized into toxic chemicals in the presence of stimulation from rays of sunlight.[28] So sunscreen doesn't just block the sunlight you need; it actually photogenerates destructive reactive oxygen species.

The aluminum commonly added to sunscreen to improve its texture and transparency and to prevent clumping of titanium dioxide is also problematic. Christopher Exley, PhD, is a chemist and an expert on aluminum whose work I mentioned in chapter 9. When Dr. Exley investigated aluminum-containing sunscreens he found that their use could result in about 200 mg of aluminum on the skin with each application.[29] WHO recommends reapplication of sunscreen every 2 hours, which means a day at the beach could result in as much as a gram of aluminum applied to the skin. Aluminum readily penetrates the skin barrier and is absorbed into the bloodstream. It is synergistically toxic with glyphosate. Both aluminum and glyphosate contribute to blocking synthesis of sulfate. Aggressive sunscreen use may be one reason for the epidemic of systemic sulfate deficiency.

There's more: Glyphosate may also be contributing to the risk of skin cancer by interfering with the supply of melanin, the natural protective pigment that produces a tan. Melanin is derived from tyrosine, a product of the shikimate pathway, which glyphosate disrupts.

Play in the Dirt!

Children don't play in the dirt as much as they used to. But making mud pies isn't just about imaginative play. Playing in the dirt enhances well-being, both for children and grown-ups. In the cultures around the world where people live hardy, healthy lives well into old age, nearly everyone has a garden. Growing your own food, coaxing flowers into bloom, and touching the soil are all healthful activities.

Both the soil microbiome and the gut microbiome have deteriorated in recent times, due to industrial-based agriculture and reduced human contact with healthy soil.[30] I feel sad that in some places it's no longer safe for children to play in soil, because you never know what heavy metals or toxic chemicals might be lurking there. Scientists believe that getting dirty was essential for the evolution of the human gut microbiome and that it's still essential for humans to thrive. Soil is an inoculant that provides a renewal of beneficial microbes. But spending too much time indoors and on screens has taken us away from frequent exposure to soil. This reduces the diversity of our microbiome, which harms our health. And, of course, the soil microbiome also suffers from pesticide exposure. Diminished soil diversity leads to reduced nutrient uptake by plants, just as a reduction in our own microbiome interferes with nutrient absorption past the gut barrier.

We need to let our kids delight in collecting earthworms, stepping in mud puddles, and playing outside. We adults need to enjoy the outdoors as well, allowing ourselves to get dirty, walk barefoot, and spend time in nature. In doing so we expose ourselves to microbes that enrich the diversity of species in the gut that lead to improved health.

Get Your Feet Back on the Ground

Getting good grounding is another easy thing to do to supply negative charge to the body, which supports good blood circulation and creates a stable internal bioelectric environment. Grounding is not just for your crazy hippie neighbor. Studies have shown that grounding, or earthing, can alleviate symptoms of chronic stress, inflammation, sleep disturbance, cardiovascular disease, and hypercoagulation of the blood.[31] The earth is a giant negatively charged ball, and when you walk barefoot on the ground, electrons move freely into your body. Especially effective is to walk barefoot in the shallow water of a sandy beach on a sunny day. The ocean air is rich in sulfur, and the water provides excellent conductance to enhance the effects of grounding. There's a reason you feel calmer, more peaceful, and more alert at the beach.

Avoid EMFs

Exposure to human-made electromagnetic fields (EMFs) has been increasing enormously. We are all being exposed to a mix of EMFs as we are bombarded by electrical charges from our appliances, smartphones, lights, television sets, and powerlines. EMF exposure is a growing concern among parents of children who now spend hours with iPads and smart phones in their laps.

Humans now use many electrically active devices all day long. These active devices generate non-natural electromagnetic fields in the surrounding air. Electromagnetic fields are created around any device that uses electricity. Even when a table lamp is not turned on, it generates an electric field in the room around it. (Magnetic fields are created only from moving electric currents, so they are generated only once the light is turned on.) High voltages are used in transmission lines, so they generate powerful electrical signals. The current that flows through them (and consequently the magnetic field strength) varies with power consumption. A cell phone generates a fairly strong electromagnetic field even when it is not being actively used. The individual frequency distribution varies from device to device, and there has been insufficient study conducted to determine if long-term exposure to these devices is damaging to our health. These nonnatural frequency disturbances may interfere with sleep. It's best to turn off the internet access in your phone, tablet, or computer at night.

We don't think of ourselves this way, but we humans are electric. There are thousands of biochemical reactions normally occurring in our bodies every second. Our neurons relay signals via electric impulses, our hearts are electrically active, and our digestion is based on biochemical reactions via the rearrangement of charged particles. This is all part of being alive. But human-made EMFs can be hazardous to our biochemistry.

Martin Pall, PhD, professor emeritus of biochemistry and basic medical sciences at Washington State University, is an expert in multiple chemical sensitivities and chronic fatigue syndrome. After over 25 years of research, Pall has demonstrated that overexposure to EMFs can harm the nervous system, the endocrine system, and even our DNA. Pall's research shows that electromagnetic fields activate voltage-gated calcium channels and induce calcium uptake into neurons, leading to neurotoxicity.[32] As we've seen, glyphosate has a similar effect, suggesting that it may be synergistically toxic with EMFs.

We can't stop the onslaught of electrical impulses coming our way. But we can take some simple steps to reduce exposure. Use an Ethernet cable at work, and get in the habit of turning off the Wi-Fi on your computer or laptop when not needed. Always put your phone in Airplane mode before going to bed. Have your children dock their electronics outside of their bedrooms. Unplug appliances when they are not in use. The microwave oven in your kitchen is also a source of EMFs, so quickly walk away after you turn it on. Even better, use the stove or a toaster oven to heat up leftovers.

Good Riddance to Bad Rubbish

"What can I do to clear glyphosate from my body?" is a question I get asked over and over again. I am saddened by how many young people and young families are suffering from avoidable health problems. I'm angered that this herbicide has been on the market for 45 years, and that we've been duped into believing it's nontoxic to humans, animals, and ecosystems. I believe we can do so many things to take control of our own health—and I've just enumerated those that I think are most important. But the truth is also that glyphosate is pervasive and nearly impossible to avoid entirely. Glyphosate needs to be banned. We need a global commitment to regenerative farming practices. And we need to attend to and nurture the land, ecosystems, and organisms that have been harmed by our shortsighted and greedy practices. I know this can be done. Let me begin with a story:

Gabe Brown had been farming for 10 years when he began working with his father-in-law on the family's 5,000-acre ranch, which had grown wheat, oats, and barley conventionally for years. After a few years, Brown and his wife purchased one-third of the ranch from her parents. For the next four years in a row, he watched his crops fail, mostly due to drought. Close to bankruptcy, Gabe cut way back on the use of herbicides, fertilizer, and pesticides. He started grazing cattle right on his fields of cash crops. The weeds accumulated.

Then something unexpected happened. Earthworms, praying mantises, and ladybugs slowly returned. The bees flew in their shaky circles around the flowering weeds, dipping their straw-like proboscises into the liquid nectar, brushing against the stigma and stamens and then rubbing some of the pollen off on the next flowering weeds. With an abundance of insects, the birds also came back.

This is how Gabe Brown discovered the effectiveness of regenerative agriculture. He started using chemical-free no-till methods with a multispecies cover crop to renew the soil during the off-season and minimize weed growth. Gabe's ranch, in Bismarck, North Dakota, now profitably produces a wide range of cash crops and cover crops, along with pastured beef, lamb, and chicken, which he sells directly to consumers.

Tainted soil cannot produce healthy plants. Under some conditions, glyphosate degrades quickly in the soil and is mostly gone within two weeks of application. However, Roundup and other glyphosate formulations last much longer in certain soil types. In fact, glyphosate and other pesticides and herbicides can be biopersistent for years. In the sandy loam soil often favored for gardening and agriculture because it retains nutrients well and allows excess water to drain away, glyphosate will persist for at least four months.[33]

How do we say good riddance to bad rubbish? Once we know that we don't want to be poisoning our planet with glyphosate, how do we get rid of it? That's a question that scientists are trying to answer. Finding new approaches to glyphosate remediation and degradation is not easy. Some of the methods being tested include electrochemical oxidation, photocatalytic degradation, hydrolytic degradation, and microbial degradation and adsorption, which takes advantage of complex molecules in the soil such as kaolinite, illite, and bentonite clay to trap glyphosate due to its negative charge, making it easier for microbial enzymes to biodegrade glyphosate.[34]

Brawny Bacteria

Microbial degradation, harnessing the power of bacteria, may be the most promising option for glyphosate remediation. Bacteria are amazingly and wonderfully adaptive, and there are a handful of microorganisms that are able to use glyphosate as a source of phosphorus, carbon, and nitrogen.[35]

Bacillus subtilis, also known as hay or grass bacillus, are rod-shaped bacteria found in the intestines of cows and other ruminants, as well as in the soil. Not only can they survive in extremely high concentrations of glyphosate, they can also process it. Under optimal fermentation conditions, *Bacillus subtilis* can degrade 65 percent of the glyphosate in their environment, even at high concentrations.[36] Critical to their success in degrading glyphosate is the presence of an isoform of the enzyme carbon-phosphorus lyase that can break the unusual carbon-phosphorus bond in glyphosate's methylphosphonyl unit. Specialized bacteria such as *Bacillus subtilis* offer a promising solution for bioremediation of highly polluted soil.

By growing cultures from soil samples taken from rice fields in eastern Nigeria, researchers managed to isolate two microbes, *Pseudomonas fluorescens* and members of the genus *Acetobacter*, that were able to use glyphosate as the source of phosphorus to grow.[37] These bacteria are able to fully biodegrade glyphosate. Whether the *Acetobacter* species present in fermented foods are also able to degrade glyphosate remains unclear; but, if this is the case, it would greatly strengthen an argument for eating fermented foods.

The Magic of Compost

Composting is an amazing process that converts organic materials such as food waste and brush into a nutrient-rich soil amendment. Composting takes advantage of bacteria to accelerate the decay process. When you build up a compost heap, much of its value lies in the humic and fulvic acids, which are complex organic molecules that possess remarkable properties. Created by microbes during the degradation of organic matter, humic and fulvic acids can act as photosensitizers, retain water, bind to clays, stimulate plant growth, and neutralize pollutants.[38] The enrichment of humic and fulvic acids in the soil can help to remove glyphosate from the soil and prevent its uptake into the plants.

Humans, too, may benefit from humic and fulvic acids to help with glyphosate detoxification. In fact, they are now promoted in supplement form.[39] It's believed these molecules bind ionically with glyphosate and then carry it out of the body through the feces. Some functional medicine specialists recommend humic and fulvic acids to help clear glyphosate in their patients.

Botulin toxins are among the most toxic agents known to biology; they work by disrupting acetylcholine receptors, causing paralysis. They are produced by the pathogen *Clostridium botulinum*. Glyphosate toxicity among dairy cows in Denmark appears to be causing chronic botulinism, probably by killing the microbes that can metabolize the toxins.[40] A study published in 2014 showed that feeding cows charcoal, sauerkraut juice, and humic acids led to a decrease in glyphosate levels in the urine and a reduction in *C. botulinum* antibodies.[41] Another study published the same year showed humic acid inhibited the deadly effects of glyphosate on beneficial bacteria, possibly by reducing its bioavailability.[42]

Wood-degrading fungi are able to degrade complex biomolecules, such as cellulose and lignin, in wood that are normally resistant to degradation. The enzymes they use to do this are powerful and general in their capabilities and have been found beneficial in clearing many synthetic chemicals, including glyphosate.[43] Examples include laccase, a phenol oxidase, and peroxidases. These enzymes work by oxidizing their substrate to a highly reactive intermediate that then breaks down the lignin or herbicide.[44]

Humic acid may contain these powerful enzymes produced by fungi. When Italian scientists studied four enzymes that are commonly found in association with humic acids—manganese peroxidase, laccase, lignin peroxidase, and horseradish peroxidase—they found that manganese peroxidase was most effective at transforming glyphosate into AMPA. Manganese peroxidase required manganese sulfate as a catalyst, and this is potentially problematic in that glyphosate chelates manganese, making it unavailable. This suggests that manganese supplementation along with humic acid is called for. Laccase was also able to degrade glyphosate, although with reduced efficacy compared to manganese peroxidase.[45] Since AMPA is also toxic, there remains the issue of disposing of the AMPA, which might be achieved by other enzymes provided by the gut microbes.

A Swedish System

Another interesting remediation strategy was developed in Sweden about 20 years ago and is now used in many European countries, Canada, and Central and South America as a way to safely dispose of pesticide waste on farms.[46] The strategy involves "biobeds," which are deep pits in the ground that feature an impermeable clay layer at the bottom, with a mixture of straw, peat, and soil above it and grass on the top. The intent of these biobeds is to provide appropriate nutrition to support bacterial and fungal microorganisms that have the capacity to degrade pesticides. Simple to install, a biobed can accelerate the degradation of environmental contaminants.

For homeowners, getting rid of Roundup stored in the garage or basement can present a serious problem. Pouring it down the drain would be a disaster, and you also shouldn't dump it into the trash. Contact your city officials or waste management department to find out where you can drop it off. Some municipalities have regularly scheduled hazardous waste disposal days, usually at the local waste transfer station. Waste management facilities usually dispose of waste materials through incineration or using industrial furnaces that can recover gases such as methane and other usable forms of energy. Or they might bury toxic waste at landfill sites. Whether the glyphosate is fully degraded through such processes, and how long it takes, is not clear.

What about Water?

While no detectable amount of glyphosate should be considered safe, the United States currently allows up to 700 micrograms of glyphosate per liter of tap water.[47] The European Union has much stricter regulations. These restrictions have motivated European researchers to study the effectiveness of various methods to remove glyphosate from contaminated water. Carbon filters are ineffective, due to glyphosate's tiny size. Unfortunately, a study investigating whether carbon filters could remove glyphosate from drinking water found that the method is effective only when the water is distilled. Water taken from the Ohio River contained organic matter that bound to the glyphosate and interfered with its filtration.[48]

Chlorine and ozone, popular methods used in municipal water treatment plants to inhibit bacterial growth, are highly oxidizing molecules that are also good at degrading glyphosate.[49] I believe this has reduced glyphosate exposure in public drinking water. Other practices, such as adding alum to the water to remove suspended solids from the water, are not effective in removing glyphosates. Treating water with ultraviolet light alone is also ineffective, but when used in combination with hydrogen peroxide, it can work well. Chlorine dioxide is a disinfectant that can kill bacteria, fungi, and viruses. Its effects on glyphosate are more variable but it can provide degradation under the right conditions. Chlorine dioxide is sometimes used instead of chlorine in water treatment plants because it is less toxic than chlorine, but this might run the risk of not adequately clearing the glyphosate.

It is never too late to change your habits in ways that can improve your health, your family's health, and the health of the planet. Let me share with you the story of Diane and Richard. Diane met the love of her life when she was well into her 70s. She was walking out of the movie theater and she noticed a man just ahead of her. He turned around and smiled. They started talking. He had bright eyes and a fetching smile, sure, but she wasn't really attracted to him. Still, there was something about his energy that Diane really liked. So, when Richard asked her to have coffee with him, she felt intrigued and agreed. And so began their romance. After two marriages that had both ended in divorce, Diane is discovering love for the first time. Their relationship is based on shared values, enthusiasm for the same important causes, and sheer enjoyment of each other. Diane feels more awake, sexually, emotionally, and spiritually, than she has in a long time.

I'm telling you about Diane and Richard's late-summer passion for two reasons. The first is that it's never too late to change something in your life, to open your eyes (and your heart) to a new idea, and to change your way of thinking. The second is that Richard, for all his lovable qualities, had a bad habit: he drank Diet Pepsi after Diet Pepsi every day. When they decided to move in together, Diane insisted that he stop drinking soda. A Wyoming boy, Richard found it hard to break the habit. But he did! Now he drinks green tea and bubbly water with lemon or lime. He hasn't had a Diet Pepsi in over a year.

Of course, the best way to cure any disease is to prevent it in the first place. Of course, it would be better if you had been eating organic your whole life, growing food in your own garden, and staying away from packaged junk food. Of course, it would be better if you had never been exposed to DDT, glyphosate, lead, aluminum, or other toxic chemicals. But if you've spent your whole life eating conventionally and ignoring your health, don't despair. Our bodies, like the Earth, tend toward healing when we support them. It's never too late to start anew.

Acknowledgments

It's impossible to list all the people who have helped me in my quest to understand glyphosate and the underlying mechanisms of biological processes in health and disease. First and foremost, I want to thank Dr. Don Huber for introducing me to glyphosate. Without that serendipitous lecture he gave in Indianapolis in September 2013, this book might not exist. Shortly after that, Joseph Mercola introduced me to Anthony Samsel, a retired toxicologist and chemist who was already keenly aware that glyphosate is much more toxic than we have been led to believe. Anthony and I collaborated over several years to jointly publish six papers on glyphosate, and he was the one who first introduced me to the idea that glyphosate might be substituting for glycine during protein synthesis. Dr. Mercola has also been a good friend and a great supporter of my work for many years.

Nancy Swanson was the first to make me aware of the fact that many modern diseases, in addition to autism, are rising dramatically in prevalence today, in line with the dramatic rise in glyphosate usage on core crops. She and I also collaborated on several papers, and Judy Hoy, a wildlife expert in Montana, joined us to help link the rise in human disease to the plight of wildlife exposed to toxic agrichemicals.

I have also collaborated with Robert Davidson, Ann Lauritzen, Glyn Wainwright, and Laurie Lentz-Marino on several joint papers where we worked through the details of how endothelial nitric oxide synthase might function as a moonlighting enzyme to switch between sulfur and nitrogen oxidation based on electromagnetic signals. Gerald Pollack helped to promote my message through invited presentations, including a TEDx Talk on sulfate's role in the body's electrical system. He is the person who invented the term "the fourth phase of water," and he has done much to popularize deep concepts in water biophysics. Others who contributed to my knowledge in this field are Mae Wan Ho and Martin Michener. I thank Glyn Wainwright for making me aware of the importance of cholesterol to mammalian biology. He and I share a common passion for the overlooked role of cholesterol sulfate in human physiology.

I thank Chen Yi Wan for inviting me in 2014 to attend a seminar in Beijing focused on the dangers of GMOs and chemical-based agriculture. It was there that I met many impassioned leaders in efforts to bring widespread awareness to the problems we face and to change our agricultural practices toward more organic and sustainable approaches. These include Xiulin Gu, Vandana Shiva, Zen Honeycutt, Bob Streit, Howard Vlieger, Monika Kruger, Judy Carman, among many others.

Sally Fallon Morell, founder of the Weston A. Price Foundation, and Teri Arranga, who organizes the yearly AutismOne conferences, have both given me forums where I have been able to present my ideas to others with shared interests. Sally has also helped shape my view on what constitutes a healthy diet.

I have met many medical practitioners who have greatly influenced my thinking and who have helped to educate me on ways to heal from glyphosate exposure. I'm in admiration of Kerri Rivera, who treats thousands of autistic children around the world, with remarkable success. Dette Avalon and Gregory Nigh share my thirst for knowledge and have both collaborated with me in writing journal articles. Joachim Gerlach has also exposed me to a vast literature on the benefits of various nutritional supplements.

Some of the many others who have touched my life and helped shape my ideas in important ways include James Beecham, David Diamond, Steve Kette, Shanhong Lu, Tony Mitra, Laura Orlando, Judy Mikovits, Luc Montagnier, Wendy Morley, and Uffe Ravnskov. A special shout-out goes to Laura Orlando, who educated me on the complexities in sewage management, and to Sarath Gunatilake, who made me aware of the crisis in kidney failure among young agricultural workers in Sri Lanka.

I am grateful to Daniel Ingersoll and Martin Michener for carefully reading early drafts of my book and suggesting important edits, and to Bethany Henrick for meticulously proofreading the science to improve the accuracy of the chapter on the gut.

It has been a pleasure to work with the entire team at Chelsea Green Publishing, including publisher Margo Baldwin, production director Patricia

Stone and her team that designed the book's beautiful cover, and copy editor Diane Durrett. A special shout-out goes to Brianne Goodspeed, associate editorial director, whose skillful editing and publishing smarts have helped bring this book to a new level of excellence.

I am also extremely grateful to Barry Lam, Chairman and Founder of Quanta Computer, Inc., and Ted Chang, its CTO. Since 2005, Quanta has been the major sponsor of my research. The topic of my research has changed over the years from natural language processing and dialog systems to human health and environmental toxicity. Their continuing belief in me and their steadfast support cannot be overstated.

Jennifer Margulis, PhD, devoted hundreds of hours of her time to helping me rework the first draft of this book so that it would be more accessible to a much larger audience. I am grateful to her for her devotion and dedication to this difficult task. Zoey O'Toole, one of the founders of Thinking Moms' Revolution, Dr. Rick Kirschner, Lindea Kirschner, Anna Houppermans, Andrew Sabatini, Peter Bockenthein, Diane Sanny, Richard Ulrich, William Parker, and James di Properzio have all helped, too. Thank you.

Finally, I want to thank my husband, Victor Zue, for his willingness to accept my preoccupation with the health of humans and the planet with so much grace and understanding. For the past several years, he has shown selfless dedication to taking care of my every need and handling the details of running our household to optimize the amount of time I could spend on this project. And he has kept me well nourished by preparing so many healthy, delicious dishes from certified organic foods. Victor, your love and support are beyond what anyone could hope for.

— APPENDIX A —

Tables

Table A.1. Twenty primary amino acids and their chemical properties.

1-Letter Code	3-Letter Code	Name	Chemistry
A	Ala	Alanine	Hydrophobic
C	Cys	Cysteine	Polar
D	Asp	Aspartic Acid	Acidic
E	Glu	Glutamic Acid	Acidic
F	Phe	Phenylalanine	Hydrophobic
G	Gly	Glycine	Hydrophobic
H	His	Histidine	Polar
I	Ile	Isoleucine	Hydrophobic
K	Lys	Lysine	Basic
L	Leu	Leucine	Hydrophobic
M	Met	Methionine	Amphipathic
N	Asn	Asparagine	Polar
P	Pro	Proline	Hydrophobic
Q	Gln	Glutamine	Polar
R	Arg	Arginine	Basic
S	Ser	Serine	Polar
T	Thr	Threonine	Polar
V	Val	Valine	Hydrophobic
W	Trp	Tryptophan	Amphipathic
Y	Tyr	Tyrosine	Amphipathic

Table A.2. Proteins involved with ATP binding that are upregulated in *E. coli* in response to glyphosate exposure.

Protein	Fold Increase
putative ATP-binding component of a transport system	2.02
D,D-dipeptide permease system, ATP-binding component	2.83
ATP-binding protein of nickel transport system	2.24
ATP-binding component of transport system for glycine, betaine, and proline	12.96
fused D-allose transporter subunits of ABC superfamily: ATP-binding components	2.03
ATP-binding component of transport system for maltose	2.38
putative ATP-binding sugar transporter	2.10
putative ATP-binding component of a transport system	3.04
putative part of putative ATP-binding component of a transport system	2.31
putative ATP-binding component of a transport system	2.30

Note: The number in the right column indicates the fold increase in protein expression. For details, see W. Lu et al. "Genome-Wide Transcriptional Responses of *Escherichia coli* to Glyphosate, a Potent Inhibitor of the Shikimate Pathway Enzyme 5-Enolpyruvylshikimate-3-Phosphate Synthase," *Molecular Biosystems* 9, no. 3 (2013): 522–30, https://doi.org/10.1039/c2mb25374g.

Table A.3. Nine proteins found to contain glyphosate substitutions that bind to phosphate-containing molecules.

Sequence	Protein Name	Phosphate Binding
AIRQTSELTLG*K	zinc finger protein 624	DNA
DG*QDRPLTKINSVK	pleckstrin homology domain-containing family A member 5	PtdIns† phosphate
EPVASLEQEEQG*K	double homeobox protein A	DNA
G*ELVMQYK	diacylglycerol kinase gamma	ATP
GKELSG*LG*SALK	very long-chain specific acyl-CoA dehydrogenase mitochondrial	FAD
KDGLG*GDK	G-protein coupled receptor 158	GTP
NEKYLG*FGTPSNLGK	ATP-dependent Clp, protease ATP-binding subunit	ATP
RTVCAKSIFELWG *HGQSPEELYSSLK	RNA (guanine(10)-N2)-methyltransferase homolog	tRNA
VTG*QLSVINSK	protein O-mannosyl-transferase 2	dolichyl phosphate

Note: G* indicates a glyphosate substitution in the peptide sequence.
† PtdIns = phosphatidyl inositol

Table A.4. Important proteins that bind heparan sulfate as a crucial part of their signaling mechanisms and to facilitate cellular clearance of debris.

	Protein	Biological Role
1	antithrombin	systemic anticoagulation
2	t-plasminogen activator	clot dissolution
3	fibroblast growth factor	stimulate mitosis (cell division)
4	interleukins and selectins	inflammation
5	ApoE	lipoprotein clearance
6	fibronectin	cell adhesion
7	laminin	cell adhesion
8	type V collagen	cell adhesion
9	thrombospondin	cell adhesion, growth

Source: Adapted from Table 29.1 in J. D. Esko, "Glycosaminoglycan-Binding Proteins," in *Essentials of Glycobiology*, eds. A. Varki et al. (Cold Spring Harbor, NY: Cold Spring Harbor Laboratory Press, 1999).

— APPENDIX B —

Recommended Resources

Organizations and Conferences

Weston A. Price Foundation (www.westonaprice.org): A nonprofit organization that promotes eating traditional, nutrient-dense foods. Weston A. Price Foundation publishes an informative quarterly magazine, hosts an annual conference, and raises awareness about the harmful effects of toxins and toxic chemicals in the food chain.

The Institute for Functional Medicine (www.ifm.org): Functional medicine addresses the root causes of diseases; and functional medical practitioners seek to heal the whole person, not just prescribe a certain medicine for a specific illness. Medical doctors who practice functional medicine are often at the cutting edge of scientific innovation, helping patients be active in improving their health.

Sustainable Agriculture Conference (www.carolinafarmstewards.org/sac): A farmer-driven membership-based nonprofit that has hosted an annual conference for over 35 years, the Carolina Farm Stewardship Association offers both practical and philosophical training for anyone interested in family farming, local agriculture, and organic food practices.

Moms Across America (www.momsacrossamerica.com): An activist organization founded by Zen Honeycutt, a mother of three, to educate and empower moms about healthy eating and healthy communities. Their website is a treasure trove of helpful information about glyphosate, detoxification, and how to become active in the movement to fix our food.

EcoFarm Conference (eco-farm.org): An annual conference sponsored by EcoFarm, an ecological farming association whose mission is to nurture safe, healthy, just, and ecologically sustainable food systems. This wonderful nonprofit, founded in 1981 and still going strong, builds alliances, fosters community events, and celebrates healthy eating and a healthy planet.

The Public Interest Network (publicinterestnetwork.org): A social change organization that has launched a campaign to ban glyphosate, The Public Interest

Network is a nationwide organization that strives to make our planet's air, food, and water safer.

Toxic-Free Future (toxicfreefuture.org): This science-forward nonprofit champions using safer products, chemicals, and practices in order to better the health of humans and the planet. Their website includes helpful basic information and scientific articles about 21 chemicals of concern.

Children's Health Defense (https://childrenshealthdefense.org): Founded by environmentalist and health advocate Robert F. Kennedy Jr., Children's Health Defense is on the forefront of the movement to protect children from toxic exposures. They champion vaccine safety, medical freedom, and government transparency.

AutismOne (www.autismone.org): This parent-centered nonprofit gives support to families affected by autism. Their annual conference includes medical doctors, researchers, and health practitioners on the cutting edge of treatment for the prevention of and recovery from autism.

Environmental Working Group (www.ewg.org): This is one of my favorite organizations. This nonprofit empowers consumers to live longer, healthier lives by sharing research and education to drive better consumer choices and nonpartisan civic action. Their motto is that everything is connected (human health and the environment), and their website contains a wealth of information about everything from contaminants in drinking water to biopersistent forever chemicals in the environment.

International Society for Environmentally Acquired Illness (iseai.org): This international nonprofit medical society raises awareness of the environmental causes of chronic diseases and seeks to help patients recover their health. You don't have to be a doctor to join, and their One People, One Health, One Planet conferences are very informative.

Environmental Health Symposium (www.environmentalhealthsymposium.com): An annual conference geared toward physicians, which is also open to the public, that highlights how environmental toxins, including pesticides, herbicides, EMFs, and GMO technology affect human health.

NutriGenic Research Institute (www.nutrigeneticresearch.org): An institute that researches nutrition and studies how the environment interfaces with epigenetics, as well as what role epigenetic factors play in human health. Their website is a good place to learn about DNA testing, mitochondrial disorders, and how single nucleotide polymorphisms may affect your health.

Organic Consumers Association (www.organicconsumers.org/usa): A grassroots nonprofit public interest group, OCA works to protect and advocate for consumers' rights to safe and healthy food. They support regenerative agriculture, corporate accountability, and organic and family farming. You can find helpful information about local farmers and where to purchase organic food on their website. They also run public interest campaigns and boycotts.

Suggestions for Further Reading

Gabe Brown, *Dirt to Soil, One Family's Journey into Regenerative Agriculture*, 2018.

Rachel Carson, *Silent Spring*, 1962.

F. William Engdahl, *Seeds of Destruction: The Hidden Agenda of Genetic Manipulation*, 2007.

Carey Gillam, *Whitewash: The Story of a Weed Killer, Cancer, and the Corruption of Science*, 2019.

Carey Gillam, *The Monsanto Papers: Deadly Secrets, Corporate Corruption, and One Man's Search for Justice*, 2021.

R. D. Lee, *Gut-Brain Secrets, Part 1: Good Food, Bad Food: (Nutrition and Toxins in Food + GMO's and Glyphosate)*, 2018.

David Perlmutter, MD, *Grain Brain: The Surprising Truth about Wheat, Carbs, and Sugar—Your Brain's Silent Killers*, 2018.

Josh Tickell, *Kiss the Ground: How the Food You Eat Can Reverse Climate Change, Heal Your Body & Ultimately Save Our World*, 2017.

E. G. Vallianatos, *Poison Spring: The Secret History of Pollution and the EPA*, 2014.

Frank A. von Hippel, *The Chemical Age: How Chemists Fought Famine and Disease, Killed Millions, and Changed Our Relationship with the Earth*, 2020.

Notes

Introduction

1. M. D. Kogan et al., "The Prevalence of Parent-Reported Autism Spectrum Disorder Among US Children," Pediatrics 142 (2018): 6, https://doi.org/10.1542/peds.2017-4161.
2. Centers for Disease Control, "Data & Statistics on Autism Spectrum Disorder," https://www.cdc.gov/ncbddd/autism/data.html.
3. Therese Limbana et al., "Gut Microbiome and Depression: How Microbes Affect the Way We Think," *Cureus* 2, no. 8 (2020) e9966, https://doi.org/10.7759/cureus.9966.
4. Hsin-Jung Wu and Eric Wu, "The Role of Gut Microbiota in Immune Homeostasis and Autoimmunity," *Gut Microbes* 3, no. 1 (2012) 4–14, https://doi.org/10.4161/gmic.19320.
5. Anastazja M. Gorecki et al., "Altered Gut Microbiome in Parkinson's Disease and the Influence of Lipopolysaccharide in a Human α-Synuclein Over-Expressing Mouse Model," *Frontiers in Neuroscience* 13 (2019) 839, https://doi.org/10.3389/fnins.2019.00839.
6. S. O. Duke et al., "Glyphosate: A Once-in-a-Century Herbicide," *Pest Management Science* 64 (2008): 319–25.
7. Molli M. Newman et al., "Changes in Rhizosphere Bacterial Gene Expression Following Glyphosate Treatment," *Science of the Total Environment* 553 (2016) 32–41, https://doi.org/10.1002/ps.1518.
8. M. R. Fernandez et al., "Glyphosate Associations with Cereal Diseases Caused by Fusarium spp. in the Canadian Prairies," *European Journal of Agronomy* 31, no. 3 (2009) 133–43, https://doi.org/10.1016/j.eja.2009.07.003.
9. Nancy L. Swanson et al., "Genetically Engineered Crops, Glyphosate and the Deterioration of Health in the United States of America," *Journal of Organic Systems* 9 (2014): 6–37.
10. "DDT," National Pesticide Information Center, 1999, http://npic.orst.edu/factsheets/ddtgen.pdf.
11. Neil Vargesson, "Thalidomide–Induced Teratogenesis: History and Mechanisms," *Birth Defects Research Part C: Embryo Today* 105, no. 2 (2015): 140–56, https://doi.org/10.1002/bdrc.21096.

12. B. Jarvis, "The Insect Apocalypse Is Here," *New York Times Magazine,* November 27, 2018, https://www.nytimes.com/2018/11/27/magazine/insect-apocalypse.html.
13. Gerardo Ceballosa et al., "Vertebrates on the Brink as Indicators of Biological Annihilation and the Sixth Mass Extinction," *Proceedings of the National Academy of Sciences* 117, no. 24 (2020): 13596–602, https://doi.org/10.1073/pnas.1704949114Corpus.

Chapter 1: Evidence of Harm

1. United States Environmental Protection Agency, "Glyphosate," https://www.epa.gov/ingredients-used-pesticide-products/glyphosate.
2. US patent number 3160632; filed: January 30, 1961; awarded: December 8, 1964.
3. US patent number 3455675 A; filed: June 25, 1968; awarded: July 15, 1969.
4. US patent number 20040077608 A1; filed: August 29, 2003; awarded: April 22, 2004.
5. Roundup. "The History of Roundup," https://www.roundup.ca/en/rounduphistory.
6. C. M. Benbrook, "Trends in Glyphosate Herbicide Use in the United States and Globally," *Environmental Sciences Europe* 28 (2016): 3, https://doi.org/10.1186/s12302-016-0070-0.
7. Charles M. Benbrook, "Trends in Glyphosate Herbicide Use in the United States and Globally," *Environmental Sciences Europe* 28 (2016): 3, https://doi.org/10.1186/s12302-016-0070-0.
8. Marie-Pier Hébert et al., "The Overlooked Impact of Rising Glyphosate Use on Phosphorus Loading in Agricultural Watersheds," *Frontiers in Ecology and the Environment 17*, no. 1 (2019): 48–56, https://doi.org/10.1002/fee.1985.
9. Narong Chamkasem and John D Vargo, "Development and Independent Laboratory Validation of an Analytical Method for the Direct Determination of Glyphosate, Glufosinate, and Aminomethylphosphonic Acid in Honey by Liquid Chromatography Tandem Mass Spectrometry," *Journal of Regulatory Science* 5, no. 2 (2017): 1–9, https://doi.org/10.21423/jrs-v05n02p001.
10. Thomas S. Thompson et al., "Determination of Glyphosate, AMPA, and Glufosinate in Honey by Online Solid-Phase Extraction-Liquid Chromatography-Tandem Mass Spectrometry," *Food Additives & Contaminants*: Part A 36, no. 2 (2019): 1–13, https://doi.org/10.1080/19440049.2019.1577993.
11. Melissa J. Perry et al., "Historical Evidence of Glyphosate Exposure from a US Agricultural Cohort," *Environmental Health* 18 (2019): 42, https://doi.org/10.1186/s12940-019-0474-6.
12. Paul J. Mills et al., "Excretion of the Herbicide Glyphosate in Older Adults between 1993 and 2016." *JAMA* 318, no. 16 (2017): 1610–11, https://doi.org/10.1001/jama.2017.11726.

13. Personal communication between Dr. Chris Chlebowski and Jennifer Margulis, July 17, 2020.
14. Monika Krüger et al., "Detection of Glyphosate Residues in Animals and Humans," *Journal of Environmental & Analytical Toxicology* 4 (2014): 210, https://doi.org/10.4172/2161-0525.1000210.
15. Sudhir Kumar et al., "Glyphosate-Rich Air Samples Induce IL-33, TSLP and Generate IL-13 Dependent Airway Inflammation," *Toxicology* 0 (2014): 42–51, https://doi.org/10.1016/j.tox.2014.08.008.
16. Carlo Caiati et al., "The Herbicide Glyphosate and Its Apparently Controversial Effect on Human Health: An Updated Clinical Perspective," *Endocrine, Metabolic & Immune Disorders—Drug Targets* 20, no. 4 (2020): 489–505, https://doi.org/0.2174/1871530319666191015191614.
17. Becky Talyn et al., "Roundup®, but Not Roundup-Ready® Corn, Increases Mortality of Drosophila melanogaster," *Toxics* 7, no. 3 (2019): 38, https://doi.org/10.3390/toxics7030038.
18. Becky Talyn et al., "Roundup®, but Not Roundup-Ready® Corn, Increases Mortality of Drosophila melanogaster," *Toxics* 7, no. 3 (2019): 38, https://doi.org/10.3390/toxics7030038.
19. Qixing Mao et al., "The Ramazzini Institute 13-Week Pilot Study on Glyphosate and Roundup Administered at Human-Equivalent Dose to Sprague Dawley Rats: Effects on the Microbiome," *Environmental Health* 17 (2018): 50, https://doi.org/10.1186/s12940-018-0394-x.
20. Jack Lewis, "Lead Poisoning: A Historical Perspective," EPA Journal. US Environmental Protection Agency, May 1985, https://archive.epa.gov/epa/aboutepa/lead-poisoning-historical-perspective.html.
21. Environmental Protection Agency, "Glyphosate,"https://www.epa.gov/ingredients-used-pesticide-products/glyphosate.
22. John Peterson Myers et al., "Concerns Over Use of Glyphosate-Based Herbicides and Risks Associated with Exposures: A Consensus Statement," *Environmental Health* 15 (2016): 19, https://doi.org/10.1186/s12940-016-0117-0.
23. European Food Safety Authority, "EFSA Explains Risk Assessment: Glyphosate," [EN] Fact sheet, https://doi.org/10.2805/654221, November 12, 2015, https://www.efsa.europa.eu/en/corporate/pub/glyphosate151112.
24. Environmental Protection Agency, "Glyphosate," https://www.epa.gov/ingredients-used-pesticide-products/glyphosate.
25. Qixing Mao et al., "The Ramazzini Institute 13-Week Pilot Study on Glyphosate and Roundup Administered at Human-Equivalent Dose to Sprague Dawley Rats: Effects on the Microbiome," *Environmental Health* 17 (2018): 50, https://doi.org/10.1186/s12940-018-0394-x.

26. Alfredo Santovito et al., "In Vitro Evaluation of Genomic Damage Induced by Glyphosate on Human Lymphocytes," *Environmental Science and Pollution Research* 25 (2018): 34693–700, https://doi.org/10.1007/s11356-018-3417-9.
27. S. Guilherme et al., "European Eel (Anguilla anguilla) Genotoxic and Pro-Oxidant Responses following Short-Term Exposure to Roundup—a Glyphosate-Based Herbicide," *Mutagenesis* 25, no. 5 (2010): 523–30, https://doi.org/10.1093/mutage/geq038; C. D. Nwani et al., "DNA Damage and Oxidative Stress Modulatory Effects of Glyphosate-Based Herbicide in Freshwater Fish, Channa punctatus," *Environ Toxicol Pharmacol* 36, no. 2 (2013): 539–47, http://dx.doi.org/10.1016/j.etap.2013.06.001.
28. M.-A. Martínez et al., "Neurotransmitter Changes in Rat Brain Regions Following Glyphosate Exposure," *Environmental Research* 161 (2018): 217, https://doi.org/10.1016/j.envres.2017.10.051.
29. Siriporn Thongprakaisang et al., "Glyphosate Induces Human Breast Cancer Cells Growth via Estrogen Receptors," *Food and Chemical Toxicology* 59 (2013): 129–36.
30. Manon Duforestel, "Glyphosate Primes Mammary Cells for Tumorigenesis by Reprogramming the Epigenome in a TET3-Dependent Manner," *Frontiers in Genetics* 10 (2019): 885, https://doi.org/10.3389/fgene.2019.00885.
31. Y. Hao et al., "Roundup® Confers Cytotoxicity through DNA Damage and Mitochondria-Associated Apoptosis Induction," *Environmental Pollution* 252, Part A (2019): 917–23, https://doi.org/10.1016/j.envpol.2019.05.128.
32. Laura N. Vandenberg et al., "Hormones and Endocrine-Disrupting Chemicals: Low-Dose Effects and Nonmonotonic Dose Responses," *Endocrine Reviews* 33(3) (2012): 378–455, https://doi.org/10.1210/er.2011-1050.
33. Eliane Dallegrave et al., "The Teratogenic Potential of the Herbicide Glyphosate—Roundup in Wistar Rats," *Toxicology Letters* 142 (2003): 45–52, https://doi.org/10.1016/s0378-4274(02)00483-6.
34. Gilles-Eric Séralini et al., "Long Term Toxicity of a Roundup Herbicide and a Roundup-Tolerant Genetically Modified Maize," *Food and Chemical Toxicology* 50 (2012); 4221–31, Retracted, https://doi.org/10.1016/j.fct.2012.08.005.
35. Claire Robinson, "Emails Reveal Role of Monsanto in Séralini Study Retraction," GMWatch, July 20, 2016, https://gmwatch.org/en/news/latest-news/17121.
36. Gilles-Eric Séralini et al., "RETRACTED: Long Term Toxicity of a Roundup Herbicide and a Roundup-Tolerant Genetically Modified Maize," *Food and Chemical Toxicology* 50, 11 (2012): 4221–31, https://www.sciencedirect.com/science/article/pii/S0278691512005637.
37. Gilles-Eric Séralini et al., "Republished Study: Long-term Toxicity of a Roundup Herbicide and a Roundup-Tolerant Genetically Modified Maize," *Environmental Sciences Europe* 26 (2014): 14, https://doi.org/10.1186/s12302-014-0014-5.

38. N. Defarge et al., "Toxicity of Formulants and Heavy Metals in Glyphosate-based Herbicides and Other Pesticides," *Toxicology Reports* 5 (2018): 156–63, https://doi.org/10.1016/j.toxrep.2017.12.025.
39. Monika Krüger et al., "Detection of Glyphosate Residues in Animals and Humans," *Journal of Environmental & Analytical Toxicology* 4 (2014): 210.
40. Awad A. Shehata et al., "Distribution of Glyphosate in Chicken Organs and Its Reduction by Humic Acid Supplementation," *The Journal of Poultry Science* 51, no. 3 (2014): 333–37, https://doi.org/https://doi.org/10.2141/jpsa.0130169.
41. Monika Krüger et al., "Detection of Glyphosate in Malformed Piglets." *Journal of Environmental and Analytical Toxicology* 4 (2014): 5, http://dx.doi.org/10.4172/2161-0525.1000230.
42. S. O. Duke et al., "Glyphosate: A Once-in-a-Century Herbicide," *Pest Management Science* 64 (2008): 319–25, https://doi.org/10.4172/2161-0525.1000210, https://doi.org/10.1002/ps.1518.
43. Daniel Brugger and Wilhelm M. Windisch, "Subclinical Zinc Deficiency Impairs Pancreatic Digestive Enzyme Activity and Digestive Capacity of Weaned Piglets," *British Journal of Nutrition* 116, no. 3 (2016): 425–33, https://doi.org/10.1017/S0007114516002105.
44. Monika Krüger et al., "Field Investigations of Glyphosate in Urine of Danish Dairy Cows," *Journal of Environmental & Analytical Toxicology* 3 (2013): 5.
45. Anthony Samsel and Stephanie Seneff, "Glyphosate, Pathways to Modern Diseases III: Manganese Neurological Diseases, and Associated Pathologies," *Surgical Neurology International* 6 (2015): 45, https://doi.org/10.4103/2152-7806.153876.
46. Channa Jayasumana et al., "Glyphosate, Hard Water and Nephrotoxic Metals: Are They the Culprits Behind the Epidemic of Chronic Kidney Disease of Unknown Etiology in Sri Lanka?" *International Journal of Environmental Research & Public Health* 11, no. 2 (2014) 2125–47, https://doi.org/10.3390/ijerph110202125.
47. Sarath Gunatilake et al., "Glyphosate's Synergistic Toxicity in Combination with Other Factors as a Cause of Chronic Kidney Disease of Unknown Origin," *International Journal of Environmental Research and Public Health* 16, no. 15 (2019): 2734, https://doi.org/10.3390/ijerph16152734.
48. Konstantin Popov et al., "Critical Evaluation of Stability Constants of Phosphonic Acids." *Pure and Applied Chemistry* 73 (2001): 1641-77, https://doi.org/10.1351/pac200173101641.
49. M. Purgel et al., "Glyphosate Complexation to Aluminium(III). An Equilibrium and Structural Study in Solution Using Potentiometry, Multinuclear NMR, ATR-FTIR, ESI-MS and DFT Calculations," *Journal of Inorganic Biochemistry* 103 (2009): 1426–38, https://doi.org/10.1016/j.jinorgbio.2009.06.011.

50. Stephanie Seneff et al., "Aluminum and Glyphosate Can Synergistically Induce Pineal Gland Pathology: Connection to Gut Dysbiosis and Neurological Disease," *Agricultural Sciences* 6 (2015): 42–70, https://doi.org/10.4236/as.2015.61005.
51. Matthew John Mold et al., "Aluminum and Neurofibrillary Tangle Co-Localization in Familial Alzheimer's Disease and Related Neurological Disorders," *Journal of Alzheimer's Disease* 78, no. 1 (2020) 139–49.
52. Katie Wedell et al., Midwest Farmers Face a Crisis. Hundreds Are Dying by Suicide, March 9, 2020, https://www.usatoday.com/in-depth/news/investigations/2020/03/09/climate-tariffs-debt-and-isolation-drive-some-farmers-suicide/4955865002/.
53. S. H. Zyoud et al., "Global Research Production in Glyphosate Intoxication from 1978 to 2015: A Bibliometric Analysis," *Human & Experimental Toxicology* 56, no. 10 (2017): 997–1006, https://doi.org/10.1177/0960327116678299.
54. A. Arul Selvi et al., "Enzyme-Linked Immunoassay for the Detection of Glyphosate in Food Samples Using Avian Antibodies," *Food and Agricultural Immunology* 22, no. 3 (2011): 217–28, https://doi.org/10.1080/09540105.2011.553799.
55. Robin Mesnage et al., "Major Pesticides Are More Toxic to Human Cells Than Their Declared Active Principles," *Biomed Research International* 2014 (2014) 179691, https://doi.org/10.1155/2014/179691.
56. C. H. Lee et al., "The Early Prognostic Factors of Glyphosate-Surfactant Intoxication," *American Journal of Emergency Medicine* 26, no. 3 (2008): 275–81, https://doi.org/10.1016/j.ajem.2007.05.011.
57. K. Zouaoui et al., "Determination of Glyphosate and AMPA in Blood and Urine from Humans: About 13 Cases of Acute Intoxication," *Forensic Science International* 226 (2013): e20–e25, https://doi.org/10.1016/j.forsciint.2012.12.010; Hsin-Ling Lee and How-Ran Guo, "The Hemodynamic Effects of the Formulation of Glyphosate-Surfactant Herbicides, in *Herbicides: Theory and Applications*," Sonia Soloneski and Marcelo L. Larramendy, eds. (London: IntechOpen, 2011), DOI: 10.5772/13486.
58. Y. H. Kim et al., "Heart Rate-Corrected QT Interval Predicts Mortality in Glyphosate-Surfactant Herbicide-Poisoned Patients," *American Journal of Emergency Medicine* 32, no. 3 (2014): 203–7, https://doi.org/10.2337/dc13-1257.
59. International Agency for Research on Cancer, "IARC Monographs Volume 112: Evaluation of Five Organophosphate Insecticides and Herbicides," March 20, 2015, *World Health Organization*, https://www.iarc.who.int/wp-content/uploads/2018/07/MonographVolume112-1.pdf.
60. As of this writing, the suit is still being contested by Bayer/Monsanto, so the final outcome is not yet clear.
61. Bloomberg, "Jury Verdict: Roundup Weed Killer was a Major Factor in Man's Cancer," *The Orange County Register,* March 19, 2019. https://www.ocregister

.com/2019/03/19/jury-verdict-roundup-weed-killer-was-a-major-factor-in-mans-cancer.

62. V. Kašuba et al., "Effects of Low Doses of Glyphosate on DNA Damage, Cell Proliferation and Oxidative Stress in the HepG2 Cell Line," *Environmental Science and Pollution Research* 24 (2017): 19267–81, https://doi.org/10.1007/s11356-017-9438-y.
63. Marta Kwiatkowska et al., "DNA Damage and Methylation Induced by Glyphosate in Human Peripheral Blood Mononuclear Cells (in Vitro Study)," *Food and Chemical Toxicology* 105 (2017): 93–98, http://dx.doi.org/10.1016/j.etap.2013.06.001.
64. J. H. Kang et al., "Methylation in the p53 Promoter is a Supplementary Route to Breast Carcinogenesis: Correlation between CpG Methylation in the p53 Promoter and the Mutation of the p53 Gene in the Progression from Ductal Carcinoma in Situ to Invasive Ductal Carcinoma," *Laboratory Investigation* 81, no. 4 (2001): 573–79, https://doi.org/10.1038/labinvest.3780266.
65. C. Paz-y-Miño et al., "Evaluation of DNA Damage in an Ecuadorian Population Exposed to Glyphosate," *Genetics and Molecular Biology* 30, no. 2 (2007): 456–60, http://dx.doi.org/10.1590/S1415-47572007000300026; C. Bolognesi et al., "Biomonitoring of Genotoxic Risk in Agricultural Workers from Five Colombian Regions: Association to Occupational Exposure to Glyphosate," *Journal of Toxicology and Environmental Health, Part A* 72, no. 15–16 (2009): 986–97, https://doi.org/10.1080/15287390902929741.
66. Mark Buchanan, "Roundup's Risks Could Go Well Beyond Cancer," *Bloomberg Opinion*, June 4, 2019, last accessed June 5, 2019, https://www.bloomberg.com/opinion/articles/2019-06-04/roundup-cancer-risk-is-only-one-danger-to-humans-animals.
67. Patricia Weiss and Tina Bellon, "Bayer Puts Roundup Future Claims Settlement on Hold," *US Legal News*, July 8, 2020, https://www.reuters.com/article/legal-us-bayer-litigation/bayer-puts-roundup-future-claims-settlement-on-hold-idUSKBN24921Q.
68. Meriel Watts et al., "Glyphosate," Pesticide Action Network International, October 2016, last accessed May 3, 2019, pan-international.org/wp-content/uploads/Glyphosate-monograph.pdf.
69. Rabah Kamal, "How Does U.S. Life Expectancy Compare to Other Countries?" Peterson-KFF Health System Tracker, December 23, 2019, https://www.healthsystemtracker.org/chart-collection/u-s-life-expectancy-compare-countries/#item-start.
70. Kenneth D. Kochanek et al., "Mortality in the United States, 2016," *NCHS Data Brief*, no. 293, December 2017.

Chapter 2: Failing Ecosystems

1. William Ophuls, Apologies to the Grandchildren, https://ophuls.org/essays.

2. Yinon M. Bar-On et al., "The Biomass Distribution on Earth," *Proceedings of the National Academy of Sciences U S A* 115, no. 25 (2018): 6506–11, https://doi.org/10.1073/pnas.1711842115.
3. Intergovernmental Science-Policy Platform on Biodiversity and Ecosystem Services, "Global Assessment Report on Biodiversity and Ecosystem Services," last accessed Oct. 10, 2010, https://www.ipbes.net/global-assessment-report-biodiversity-ecosystem-services.
4. J. E. Primost et al., "Glyphosate and AMPA, 'Pseudo-Persistent' Pollutants Under Real-World Agricultural Management Practices in the Mesopotamic Pampas Agroecosystem, Argentina," *Environmental Pollution* 229 (2017): 771–77, https://doi.org/10.1016/j.envpol.2017.06.006.
5. P. Laitinen et al., "Fate of the Herbicide Glyphosate, Glyphosinate-Ammonium, Phenmedipham, Ethofumesate and Metamitron in Two Finnish Arable Soils," *Pest Management Science* 62 (2006): 473–91, https://doi.org/10.1002/ps.1186.
6. J. E. Primost et al., "Glyphosate and AMPA, 'Pseudo-Persistent' Pollutants Under Real-World Agricultural Management Practices in the Mesopotamic Pampas Agroecosystem, Argentina," *Environmental Pollution* 229 (2017): 771–77, https://doi.org/10.1016/j.envpol.2017.06.006.
7. Lars Bergström et al., "Laboratory and Lysimeter Studies of Glyphosate and Aminomethylphosphonic Acid in a Sand and a Clay Soil," *Journal of Environmental Quality* 40 (2011): 98–108, https://doi.org/10.2134/jeq2010.0179.
8. Florence Poirier et al., "Proteomic Analysis of the Soil Filamentous Fungus Aspergillus nidulans Exposed to a Roundup Formulation at a Dose Causing No Macroscopic Effect: A Functional Study," *Environmental Science and Pollution Research International* 24, no. 33 (2017): 25933–46, https://doi.org/0.1007/s11356-017-0217-6.
9. Jacqueline E. Mohan, "Mycorrhizal Fungi Mediation of Terrestrial Ecosystem Responses to Global Change: Mini-Review," *Fungal Ecology* 10 (2014) 3–19, https://doi.org/10.1016/j.funeco.2014.01.005.
10. Nic Fleming, "Plants Talk to Each Other Using an Internet of Fungus," November 11, 2014, http://www.bbc.com/earth/story/20141111-plants-have-a-hidden-internet.
11. G. Murray and J. Brennan, "Estimating Disease Losses to the Australian Wheat Industry," *Australasian Plant Pathology* 38, no. 6 (2009) 558–70, https://doi.org/10.1071/AP09064.
12. M. R. Fernandez et al., "Glyphosate Associations with Cereal Diseases Caused by Fusarium spp. in the Canadian Prairies," *European Journal of Agronomy* 31, no. 3 (2009) 133–43, https://doi.org/10.1016/j.eja.2009.07.003.

13. Martin V. Dutton and Christine S. Evans, "Oxalate Production by Fungi: Its Role in Pathogenicity and Ecology in the Soil Environment," *Canadian Journal of Microbiology* 42 (1996): 881–95, https://doi.org/10.1139/m96-114.
14. A. Casadevall, "Fungal Diseases in the 21st Century: The Near and Far Horizons," *Pathogens and Immunity* 3 (2018): 183–96, https://doi.org/10.20411/pai.v3i2.249.
15. Felix Bongomin et al., "Global and Multi-National Prevalence of Fungal Diseases—Estimate Precision," *Journal of Fungi (Basel)*. 3, no. 4 (2017) 57, https://doi.org/10.3390/jof3040057; Fausto Almeida et al., "The Still Underestimated Problem of Fungal Diseases Worldwide," *Frontiers in Microbiology* 10 (2019): 214, https://doi.org/10.3389/fmicb.2019.00214.
16. Katharina Kainz et al., "Fungal Infections: The Hidden Crisis," *Microbial Cell* 7, no. 6 (2020) 143–45, http://doi.org/10.15698/mic2020.06.718.
17. Fungal Diseases, Centers for Disease Control and Prevention, "Types of Fungal Diseases," https://www.cdc.gov/fungal/diseases/index.html.
18. C. S. Carranza et al., "Glyphosate in Vitro Removal and Tolerance by Aspergillus oryzae in Soil Microcosms," *International Journal of Environmental Science and Technology* 16 (2019) 7673–82, https://doi.org/10.1007/s13762-019-02347-x.
19. M. Cristina Romero et al., "Biodegradation of Glyphosate by Wild Yeasts," *Revista Mexicana de Micología* 19 (2004) 46–50.
20. M. A. Pfaller et al., "*Candida krusei*, a Multidrug-Resistant Opportunistic Fungal Pathogen: Geographic and Temporal Trends from the ARTEMIS DISK Antifungal Surveillance Program, 2001 to 2005," *Journal of Clinical Microbiology* 46, no. 2 (2008) 515–21, https://doi.org/10.1128/JCM.01915-07.
21. Cornelius J. Clancy, "Why the CDC Warns Antibiotic-Resistant Fungal Infections are an Urgent Health Threat," The Conversation, November 19, 2019, https://theconversation.com/why-the-cdc-warns-antibiotic-resistant-fungal-infections-are-an-urgent-health-threat-127095.
22. Centers for Disease Control, "Tracking *Candida auris*," last accessed November 25, 2020, https://www.cdc.gov/fungal/candida-auris/tracking-c-auris.html.
23. Anna Jeffery-Smith et al., "*Candida auris*: A Review of the Literature," *Clinical Microbiology Reviews* 31 (2017): e00029–17, https://doi.org/10.1128/CMR.00029-17.
24. W. A. Battaglin et al., "Glyphosate and Its Degradation Product AMPA Occur Frequently and Widely in U.S. Soils, Surface Water, Groundwater, and Precipitation," *Journal of the American Water Resources Association* 50, no. 2 (2014): 275–90, https://doi.org/10.1111/jawr.12159.
25. Juan Manuel Montiel-León et al., "Widespread Occurrence and Spatial Distribution of Glyphosate, Atrazine, and Neonicotinoids Pesticides in the St. Lawrence

and Tributary Rivers," *Environmental Pollution* 250 (2019): 29–39, https://doi.org/10.1016/j.envpol.2019.03.125.

26. Rocío Inés Bonansea et al., "The Fate of Glyphosate and AMPA in a Freshwater Endorheic Basin: An Ecotoxicological Risk Assessment," *Toxics* 6, no. 3 (2018), http://doi.org/10.3390/toxics6010003.
27. J. L. Stoddard et al., "Continental-Scale Increase in Lake and Stream Phosphorus: are Oligotrophic Systems Disappearing in the United States?" *Environmental Science and Technology* 50 (2016): 3409–15, https://doi.org/10.1021/acs.est.5b05950.
28. Kevin W. King et al., "Contributions of Systematic Tile Drainage to Watershed-Scale Phosphorus Transport," *Journal of Environmental Quality* 44, no. 2 (2015): 486–94, https://doi.org/10.2134/jeq2014.04.0149.
29. Kevin W. King et al., "Contributions of Systematic Tile Drainage to Watershed-Scale Phosphorus Transport," *Journal of Environmental Quality* 44, no. 2 (2015): 486–94, https://doi.org/10.2134/jeq2014.04.0149.
30. Damian Drzyzga and Jacek Lipok, "Glyphosate Dose Modulates the Uptake of Inorganic Phosphate by Freshwater Cyanobacteria," *Journal of Applied Phycology* 30 (2018): 299–309, https://doi.org/10.1007/s10811-017-1231-2.
31. Philip Mercurio et al., "Glyphosate Persistence in Seawater," *Marine Pollution Bulletin* 85 (2014): 385–90, https://doi.org/10.1016/j.marpolbul.2014.01.021.
32. Antonio Suppa et al., "Roundup Causes Embryonic Development Failure and Alters Metabolic Pathways and Gut Microbiota Functionality in Non-target Species," *Microbiome* 8 (2020): 170, https://doi.org/10.1186/s40168-020-00943-5.
33. I. S. Canosa et al., "Imbalances in the Male Reproductive Function of the Estuarine Crab *Neohelice granulata*, Caused by Glyphosate," *Ecotoxicology and Environmental Safety* 182 (2019): 109405, https://doi.org/10.1016/j.ecoenv.2019.109405.
34. Mahdi Banaee et al., "Acute Exposure to Chlorpyrifos and Glyphosate Induces Changes in Hemolymph Biochemical Parameters in the Crayfish, *Astacus leptodactylus* (Eschscholtz, 1823)," *Comparative Biochemistry and Physiology, Part C* 222 (2019): 145–55, https://doi.org/10.1016/j.cbpc.2019.05.003.
35. S. D. Wang et al., "Inhibitory Effects of 4-Dodecylresorcinol on the Phenoloxidase of the Diamondback Moth *Plutella xylostella* (L.) (Lepidoptera Plutellidae)," *Pesticide Biochemistry and Physiology* 82, no. 1 (2005): 52–58, https://doi.org/10.1016/j.pestbp.2004.12.005.
36. Mahdi Banaee et al., "Acute Exposure to Chlorpyrifos and Glyphosate Induces Changes in Hemolymph Biochemical Parameters in the Crayfish, *Astacus leptodactylus* (Eschscholtz, 1823)," *Comparative Biochemistry and Physiology, Part C* 222 (2019): 145–55, https://doi.org/10.1016/j.cbpc.2019.05.003.
37. Mirko Manchia et al., "Serotonin Dysfunction, Aggressive Behavior, and Mental Illness: Exploring the Link Using a Dimensional Approach," *ACS Chemical Neuroscience* 8, no. 5 (2017): 961–72, https://doi.org/10.1021/acschemneuro.6b00427.

38. David A. Davis et al., "Cyanobacterial Neurotoxin BMAA and Brain Pathology in Stranded Dolphins," *PLoS ONE* 14, no. 3 (2019): e0213346, https://doi.org/10.1371/journal.pone.0213346.
39. Francisco Sánchez-Bayo and Kris A. G. Wyckhuys, "Worldwide Decline of the Entomofauna: A Review of Its Drivers," *Biological Conservation* 232 (2019): 8–27, https://doi.org/10.1016/j.biocon.2019.01.020.
40. Caspar A. Hallmann et al., "More than 75 Percent Decline over 27 Years in Total Flying Insect Biomass in Protected Areas," *PLoS ONE* 12, no. 10 (2017): e0185809, https://doi.org/10.1371/journal.pone.0185809.
41. Noel Kirkpatrick, "California's Monarch Butterfly Population Has Declined by 99% since the 1980s," Treehugger, January 30, 2020, https://www.treehugger.com/california-monarch-butterfly-population-drops-single-year-4861288.
42. Sarah P. Saunders et al., "Local and Cross-Seasonal Associations of Climate and Land Use with Abundance of Monarch Butterflies *Danaus plexippus*," *Ecography* 41, no. 2 (2018): 278–90, https://doi.org/10.1111/ecog.02719.
43. Marek Cuhra et al., "Glyphosate-Residues in Roundup-Ready Soybean Impair *Daphnia magna* Life-Cycle," *Journal of Agricultural Chemistry and Environment* 4 (2015): 24–36, https://doi.org/10.4236/jacen.2015.41003.
44. S. G. Potts et al., "Global Pollinator Declines: Trends, Impacts and Drivers," *Trends in Ecology and Evolution* 25 (2010): 345–53, https://doi.org/10.1016/j.tree.2010.01.007.
45. A. Fairbrother et al., "Risks of Neonicotinoid Insecticides to Honeybees," *Environmental Toxicology and Chemistry* 33 (2018): 719–31, https://doi.org/10.1002/etc.2527.
46. Kamila Derecka et al., "Transient Exposure to Low Levels of Insecticide Affects Metabolic Networks of Honeybee Larvae," *PLoS ONE* 8, no. 7 (2013): e68191, https://doi.org/10.1371/journal.pone.0068191.
47. E. Hetanen et al., "Effects of Phenoxyherbicides and Glyphosate on the Hepatic and Intestinal Biotransformation Activities in the Rat," *Acta Pharmacologica et Toxicologica* 53 (1983): 103–12, https://doi.org/10.1111/j.1600-0773.1983.tb01876.x; Mohamed Ahmed Fathi et al., "Disruption of Cytochrome P450 Enzymes in the Liver and Small Intestine in Chicken Embryos in Ovo Exposed to Glyphosate," *Environmental Science and Pollution Research* 27, no. 14 (2020): 16865–75.
48. Annette McGivney, "'Like Sending Bees to War': The Deadly Truth behind Your Almond Milk Obsession," The Guardian, January 8, 202, https://www.theguardian.com/environment/2020/jan/07/honeybees-deaths-almonds-hives-aoe.
49. Lucila T. Herbert et al., "Effects of Field-Realistic Doses of Glyphosate on Honeybee Appetitive Behavior," *Experimental Biology* 1, no. 217 (Pt 19) (2014): 3457–64, https://doi.org/10.1242/jeb.109520.

50. Mara Sol Balbuena et al., "Effects of Sublethal Doses of Glyphosate on Honeybee Navigation," *Journal of Experimental Biology* 218 (2015): 2799–805, https://doi.org/10.1242/jeb.117291.
51. Walter M. Farina et al., "Effects of the Herbicide Glyphosate on Honey Bee Sensory and Cognitive Abilities: Individual Impairments with Implications for the Hive," *Insects* 10 (2019): 354, https://doi.org/10.3390/insects10100354.
52. E. V. S . Motta et al., "Glyphosate Perturbs the Gut Microbiota of Honey Bees," *Proceedings of the National Academy of Sciences of the United States of America* 115 (2018): 10305–10, https://doi.org/10.1073/pnas.1803880115; Nicolas Blot et al., "Glyphosate, but Not Its Metabolite AMPA, Alters the Honeybee Gut Microbiota," *PLoS ONE* 14, no. 4 (2019): e0215466, https://doi.org/10.1371/journal.pone.0215466.
53. Dominic Martella, "Tillage Farming Damaging Earthworm Populations," *Science Daily*, May 8, 2017, https://www.sciencedaily.com/releases/2017/05/170508095152.htm.
54. Jacqueline L. Stroud, "Soil Health Pilot Study in England: Outcomes from an On-Farm Earthworm Survey," *PLoS ONE* 14, no. 2 (2019): e0203909; https://doi.org/10.1371/journal.pone.0203909.
55. Rothamsted Research, "Earthworm Research Spurs Farmers to Act," February 21, 2019, https://www.rothamsted.ac.uk/news/earthworm-research-spurs-farmers-act.
56. Mailin Gaupp-Berghausen et al., "Glyphosate-Based Herbicides Reduce the Activity and Reproduction of Earthworms and Lead to Increased Soil Nutrient Concentrations," *Scientific Reports* 5 (2015): 12886, https://doi.org/10.1038/srep12886.
57. Specifically, myosin and PEP carboxykinase (PEPCK), which I talk more about in chapter 7.
58. Rekek Negga et al. "Exposure to Glyphosate- and/or Mn/Zn-Ethylene-bis-Dithiocarbamate-Containing Pesticides Leads to Degeneration of γ-Aminobutyric Acid and Dopamine Neurons in *Caenorhabditis elegans*," *Neurotoxicity Research* 21," no. 3 (2012): 281–90, https://doi.org/10.1007/s12640-011-9274-7.
59. Ben C. Scheele et al., "Amphibian Fungal Panzootic Causes Catastrophic and Ongoing Loss of Biodiversity," *Science* 363, no. 6434 (2019): 1459–63, https://doi.org/10.1126/science.aav0379.
60. R. Monastersky, "Life—A Status Report," *Nature* 516 (2014): 158–61.
61. Rick A. Relyea, "The Lethal Impact of Roundup on Aquatic and Terrestrial Amphibians," *Ecological Applications* 15, no. 4 (2005): 1118–24, https://doi.org/10.1890/04-1291.
62. Sylvain Slaby et al., "Effects of Glyphosate and a Commercial Formulation Roundup® Exposures on Maturation of Xenopus laevis Oocytes," *Environmental Science and Pollution Research* 27, no. 4 (2020):3697–705, https://doi.org/10.1007/s11356-019-04596-2.

63. Rafael C. Lajmanovich et al., "First Evaluation of Novel Potential Synergistic Effects of Glyphosate and Arsenic Mixture on *Rhinella arenarum* (Anura: Bufonidae) Tadpoles." *Heliyon* 5 (2019): e02601, https://doi.org/10.1016/j.heliyon.2019.e02601.
64. Rajendiran Karthkraj and Kurunthachalam Kannan, "Widespread Occurrence of Glyphosate in Urine from Pet Dogs and Cats in New York State, USA," *Science of The Total Environment* 659 (2019): 790–95, https://doi.org/10.1016/j.scitotenv.2018.12.454.
65. Amjad P. Khan et al., "The Role of Sarcosine Metabolism in Prostate Cancer Progression," *Neoplasia* 15, no. 5 (2013): 491–501, https://doi.org/10.1593/neo.13314.
66. Lisa M. Freeman et al., "Diet-Associated Dilated Cardiomyopathy in Dogs: What Do We Know?" *Journal of the American Veterinary Medical Association* 253, no. 11 (2018): 1390–94, https://doi.org/10.2460/javma.253.11.1390.
67. W. Lu et al., "Genome-Wide Transcriptional Responses of *Escherichia coli* to Glyphosate, a Potent Inhibitor of the Shikimate Pathway Enzyme 5-Enolpyruvylshikimate-3-Phosphate Synthase," *Molecular Biosystems* 9, no. 3 (2013): 522–30, https://doi.org/10.1039/c2mb25374g.
68. GMO Free USA versus Nestlé Purina Petcare Company, Case No. 2020 CA 002775 B, June 19, 2020, https://gmofreeusa.org/wp-content/uploads/2020/06/Purina-ComplaintPacket_GMOToxinFreeUSA_CleanLabelProject_20200618.pdf.
69. GMO Free USA, "Ethoxyquin: Is There Something Fishy about Your Pet Food?" https://gmofreeusa.org/food-testing/ethoxyquin-is-there-something-fishy-about-your-pet-food.

Chapter 3: Glyphosate and the Microbiome

1. Nancy L. Swanson et al., "Genetically Engineered Crops, Glyphosate and the Deterioration of Health in the United States of America," *Journal of Organic Systems* 9 (2014): 6–37.
2. Nancy L. Swanson et al., "Genetically Engineered Crops, Glyphosate and the Deterioration of Health in the United States of America," *Journal of Organic Systems* 9 (2014): 6–37.
3. Lyydia Leino et al., "Classification of the Glyphosate Target Enzyme (5-Enolpyruvylshikimate-3-Phosphate Synthase) for Assessing Sensitivity of Organisms to the Herbicide," *Journal of Hazardous Materials* 2020 (Epub ahead of print); https://doi.org/10.1016/j.jhazmat.2020.124556.
4. Stefanía Magnúsdóttir et al. "Systematic Genome Assessment of B-Vitamin Biosynthesis Suggests Co-operation among Gut Microbes," *Frontiers in Genetics* 6 (2015): 148, https://doi.org/10.3389/fgene.2015.00148.
5. Ron Sender et al., "Revised Estimates for the Number of Human and Bacteria Cells in the Body," *PLoS Biology* 14(8) (2016): e1002533, https://doi.org/10.1371/journal.pbio.1002533.

6. C. A. Lozupone et al., "Diversity, Stability and Resilience of the Human Gut Microbiota," *Nature* 489 (2012): 220–30, https://doi.org/10.1038/nature11550.
7. Claudio Cristiano et al., "Interplay between Peripheral and Central Inflammation in Autism Spectrum Disorders: Possible Nutritional and Therapeutic Strategies," *Frontiers in Physiology* 9 (2018) 184, https://doi.org/10.3389/fphys.2018.00184.
8. Clair R. Martin, "The Brain-Gut-Microbiome Axis," *Cellular and Molecular Gastroenterology and Hepatology* 6, no. 2 (2018): 133–48, https://doi.org/10.1016/j.jcmgh.2018.04.003.
9. Esther E. Fröhlich et al., "Cognitive Impairment by Antibiotic-Induced Gut Dysbiosis: Analysis of Gut Microbiota-Brain Communication," *Brain, Behavior, and Immunity* 56 (2016) 140–55, https://doi.org/10.1016/j.bbi.2016.02.020.
10. Luisa Möhle et al., "Ly6Chi Monocytes Provide a Link between Antibiotic-Induced Changes in Gut Microbiota and Adult Hippocampal Neurogenesis," *Cell Reports* 15 (2016): 1945–56, https://doi.org/10.1016/j.celrep.2016.04.074.
11. Yan Shao et al., "Stunted Microbiota and Opportunistic Pathogen Colonization in Caesarean-Section Birth." *Nature* 574 (2019): 117–21, https://doi.org/10.1038/s41586-019-1560-1.
12. Gonzalo N. Bidart et al., "The Lactose Operon from *Lactobacillus casei* is Involved in the Transport and Metabolism of the Human Milk Oligosaccharide Core-2 N-acetyllactosamine," *Scientific Reports* 8 (2018): 7152. https://doi.org/10.1038/s41598-018-25660-w.
13. Claudio Cristiano et al., "Interplay between Peripheral and Central Inflammation in Autism Spectrum Disorders: Possible Nutritional and Therapeutic Strategies," *Frontiers in Physiology* 9 (2018) 184, https://doi.org/10.3389/fphys.2018.00184.
14. Microbial metabolites such as lipopolysaccharide (LPS) and phenolic compounds such as *p*-cresol.
15. Isadora Argou-Cardozo and Fares Zeidán-Chuliá, "*Clostridium* Bacteria and Autism Spectrum Conditions: A Systematic Review and Hypothetical Contribution of Environmental Glyphosate Levels," *Medical Sciences (Basel)* 6, no. 2 (2018): 29, https://doi.org/10.3390/medsci6020029.
16. Elaine Y. Hsiao et al., "Microbiota Modulate Behavioral and Physiological Abnormalities Associated with Neurodevelopmental Disorders," *Cell* 155 (2013): 1451–63, https://doi.org/10.1016/j.cell.2013.11.024.
17. A. M. Persico and V. Napolioni, "Urinary p-Cresol in Autism Spectrum Disorder," *Neurotoxicology & Teratology* 36 (2013): 82–90, https://doi.org/10.1016/j.ntt.2012.09.002.
18. Kailyn L. Stefan et al., "Commensal Microbiota Modulation of Natural Resistance to Virus Infection," *Cell* 103, no. 5 (2020): 1312–24, https://doi.org/10.1016/j.cell.2020.10.047.

19. N. Sudo et al., "Postnatal Microbial Colonization Programs the Hypothalamic-Pituitary-Adrenal System for Stress Response in Mice," *Journal of Physiology* 558 (2004): 263–75, https://doi.org/10.1113/jphysiol.2004.063388.
20. Angela Vince et al., "Ammonia Production by Intestinal Bacteria," *Gut* 14, no. 3 (1973): 171–77, https://doi.org/10.1136/gut.14.3.171.
21. A. A. Shehata et al., "The Effect of Glyphosate on Potential Pathogens and Beneficial Members of Poultry Microbiota *In Vitro*," *Current Microbiology* 66, no. 4 (2013): 350–58, https://doi.org/10.1007/s00284-012-0277-2.
22. Lene Norby Nielsen et al., "Glyphosate Has Limited Short-Term Effects on Commensal Bacterial Community Composition in the Gut Environment due to Sufficient Aromatic Amino Acid Levels," *Environmental Pollution* 233 (2018): 364e376, https://doi.org/10.1016/j.envpol.2017.10.016.
23. Anthony Samsel and Stephanie Seneff, "Glyphosate Pathways to Modern Diseases VI: Prions, Amyloidoses and Autoimmune Neurological Diseases," *Journal of Biological Physics & Chemistry* 17 (2017): 8–32, https://doi.org/10.4024/25SA16A.jbpc.17.01.
24. Harold L. Newmark et al., "Determinants and Consequences of Colonic Luminal pH: Implications for Colon Cancer," *Nutrition and Cancer* 14, no. 3–4 (1990): 161–73, https://doi.org/10.1080/01635589009514091.
25. Samuel K. Lai, "Micro- and Macrorheology of Mucus," *Advanced Drug Delivery Reviews* 61, no. 2 (2009): 86–100, https://doi.org/10.1016/j.addr.2008.09.012.
26. Jung-hyun Rho et al., "A Novel Mechanism for Desulfation of Mucin: Identification and Cloning of a Mucin-Desulfating Glycosidase (Sulfoglycosidase)," *Journal of Bacteriology* 187, no. 5 (2005): 1543–51, https://doi.org/10.1128/JB.187.5.1543-1551.2005.
27. Nicola Volpi et al., "Human Milk Glycosaminoglycan Composition from Women of Different Countries: A Pilot Study," *The Journal of Maternal-Fetal & Neonatal Medicine* (2018): 1–61, https://doi.org/10.1080/14767058.2018.1539309; G. V. Coppa et al., "Composition and Structure Elucidation of Human Milk Glycosaminoglycans," *Glycobiology* 21, no.3 (2011): 295—303, https://doi.org/10.1093/glycob/cwq164.
28. Bethany M. Henric et al., "Elevated Fecal pH Indicates a Profound Change in the Breastfed Infant Gut Microbiome Due to Reduction of Bifidobacterium over the Past Century," *mSphere* 3, no. 2 (2018): e00041–18, https://doi.org/10.1128/mSphere.00041-18.
29. Giorgio Casaburi et al., "Metagenomic Insights of the Infant Microbiome Community Structure and Function across Multiple Sites in the United States," Scientific Reports 11 (2021): 1472, https://doi.org/10.1038/s41598-020-80583-9.

30. W. R. Logan, "The Intestinal Flora of Infants and Young Children," *Journal of Pathology* 18 (1913): 527–51. https://doi.org/10.1002/path.1700180154.
31. S. A. Frese et al., "Persistence of Supplemented *Bifidobacterium longum* subsp. *infantis* EVC001 in Breastfed Infants," *mSphere* 2 (2017): e00501–17, https://doi.org/10.1128/mSphere.00501-17.
32. A. A. Shehata et al., "The Effect of Glyphosate on Potential Pathogens and Beneficial Members of Poultry Microbiota *In Vitro*," *Current Microbiology* 66, no. 4 (2013): 350–58, https://doi.org/10.1007/s00284-012-0277-2.
33. Maria M. Milesi et al., "Perinatal Exposure to a Glyphosate-Based Herbicide Impairs Female Reproductive Outcomes and Induces Second-Generation Adverse Effects in Wistar Rats," *Archives of Toxicology* 92, no. 8 (2018): 2629–43, https://doi.org/10.1007/s00204-018-2236-6; Deepika Kubsad et al., "Assessment of Glyphosate Induced Epigenetic Transgenerational Inheritance of Pathologies and Sperm Epimutations: Generational Toxicology," *Scientific Reports* 9 (2019): 6372, https://doi.org/10.1038/s41598-019-42860-0.
34. S. Parvez et al., "Glyphosate Exposure in Pregnancy and Shortened Gestational Length: A Prospective Indiana Birth Cohort Study," *Environmental Health* 17 (2018): 23, https://doi.org/10.1186/s12940-018-0367-0.
35. Bethany M. Henrick et al.,"Colonization by *B. infantis* EVC001 Modulates Enteric Inflammation in Exclusively Breastfed Infants," *Pediatric Research* 86 (2019): 7490-757, https://doi.org/10.1038/s41390-019-0533-2.
36. Bethany M. Henrick et al., "Elevated Fecal pH Indicates a Profound Change in the Breastfed Infant Gut Microbiome Due to Reduction of *Bifidobacterium* over the Past Century," *mSphere* 3, no. 2 (2018): e00041–18, https://doi.org/10.1128/mSphere.00041-18.
37. Nadia Regina Rodrigues and Ana Paula Ferreira de Souza. "Occurrence of Glyphosate and AMPA Residues in Soy-Based Infant Formula Sold in Brazil." *Food Additives & Contaminants: Part A* 35, no. 4 (2018): 724–31, https://doi.org/10.1080/19440049.2017.1419286.
38. M. A. Priestman et al., "5-Enolpyruvylshikimate-3-Phosphate Synthase from *Staphylococcus aureus* Is Insensitive to Glyphosate," *FEBS Letters* 579 (2005): 728–32, https://doi.org/10.1016/j.febslet.2004.12.057.
39. Dennis Wicke et al., "Identification of the First Glyphosate Transporter by Genomic Adaptation," *Environmental Microbiology* 21, no. 4 (2019): 1287–305, https://doi.org/10.1111/1462-2920.14534.
40. Dilip Nathwani et al., "Clinical and Economic Consequences of Hospital-Acquired Resistant and Multidrug-Resistant *Pseudomonas aeruginosa* Infections: A Systematic Review and Meta-Analysis," *Antimicrobial Resistance and Infection Control* 3 (2014): 32, https://doi.org/10.1186/2047-2994-3-32.

41. Robert Gaynes, "The Discovery of Penicillin—New Insights after More Than 75 Years of Clinical Use," *Emerging Infectious Diseases* 23, no. 5 (2017) 849–53, https://doi.org/10.3201/eid2305.161556, https://www.ncbi.nlm.nih.gov/pmc/articles/PMC5403050/.
42. Centers for Disease Control and Prevention, "Antibiotic/Antimicrobial Resistance (AR/AMR," https://www.cdc.gov/drugresistance/biggest-threats.html.
43. Brigitta Kurenbach et al., "Sublethal Exposure to Commercial Formulations of the Herbicides Dicamba, 2,4-Dichlorophenoxyacetic Acid, and Glyphosate Cause Changes in Antibiotic Susceptibility in *Escherichia coli* and *Salmonella enterica serovar Typhimurium*," *mBio* 6, no. 2 (2015): e00009–15, https://doi.org/10.1128/mBio.00009-15.
44. Konrad C. Bradley et al., "Microbiota-Driven Tonic Interferon Signals in Lung Stromal Cells Protect from Influenza Virus Infection," *Cell Reports* 28 (2019): 245–56, https://doi.org/10.1016/j.celrep.2019.05.105.
45. David Ríos-Covián et al., "Intestinal Short-Chain Fatty Acids and Their Link with Diet and Human Health," *Frontiers in Microbiology* 7 (2016): 185, https://doi.org/10.3389/fmicb.2016.00185.
46. Jeonghyun Choi et al., "Pathophysiological and Neurobehavioral Characteristics of a Propionic Acid-Mediated Autism-Like Rat Model," *PLoS ONE* 13, no. 2 (2018): e0192925, https://doi.org/10.1371/journal.pone.0192925.
47. Alan W. Walker et al., "pH and Peptide Supply Can Radically Alter Bacterial Populations and Short-Chain Fatty Acid Ratios within Microbial Communities from the Human Colon," *Applied and Environmental Microbiology* (2005): 3692–700, https://doi.org/10.1128/AEM.71.7.3692-3700.2005.
48. This effect was significant, with a *p*-value of 0.001 (0.05 is the threshold for significance).
49. I. Mangin et al., "Molecular Inventory of Faecal Microflora in Patients with Crohn's Disease," *FEMS Microbiology Ecology* 50, no. 1 (2004): 25–36, https://doi.org/10.1016/j.femsec.2004.05.005.
50. B. T, Welch et al., "Auto-brewery Syndrome in the Setting of Long-standing Crohns Disease: A Case Report and Review of the Literature," *Journal of Crohn's and Colitis* (2016): 1448–50, https://doi.org/10.1093/ecco-jcc/jjw098.
51. Kelly Painter and Kristin L. Sticco, "Auto-brewery Syndrome (Gut Fermentation)" (Treasure Island, FL: StatPearls).
52. Carol A. Kumamoto et al., "Inflammation and Gastrointestinal Candida Colonization," *Current Opinion in Microbiology* 14, no. 4 (2011): 386–91, https://doi.org/10.1016/j.mib.2011.07.015.
53. M. A. Pfaller and D. J. Diekema, "Epidemiology of Invasive Candidiasis: A Persistent Public Health Problem," *Clinical Microbiology Reviews* 20, no. 1 (2007): 133–63, https://doi.org/10.1128/CMR.00029-06.

54. M. M, McNeil et al., "Trends in Mortality due to Invasive Mycotic Diseases in the United States, 1980–1997," *Clinical Infectious Diseases* 33 (2001): 641–47, https://doi.org/10.1086/322606.
55. G. S. Martin et al., "The Epidemiology of Sepsis in the United States from 1979 through 2000," *New England Journal of Medicine* 348 (2003): 1546–54, https://doi.org/10.1056/NEJMoa022139.
56. Amnon Sonnenberg et al., "Epidemiology of Constipation in the United States," *Diseases of the Colon & Rectum* 32, no. 1 (1989) 1–8, https://doi.org/10.1007/BF02554713.
57. Sun Jung Oh et al, "Chronic Constipation in the United States: Results from a Population-Based Survey Assessing Healthcare Seeking and Use of Pharmacotherapy," The American Journal of Gastroenterology 115, no. 6 (2020) 895–905, https://doi.org/10.14309/ajg.0000000000000614.
58. T. Sommers et al., "Emergency Department Burden of Constipation in the United States from 2006 to 2011," *American Journal of Gastroenterology* 110, no. 4 (2015): 572–79, https://doi.org/10.1038/ajg.2015.64.
59. Hajime Nakae et al., "Paralytic Ileus Induced by Glyphosate Intoxication Successfully Treated Using Kampo Medicine," *Acute Medicine & Surgery* 2, no. 3 (2015): 214–18, https://doi.org/10.1002/ams2.103.
60. Lu Fan et al., "Glyphosate Effects on Symbiotic Nitrogen Fixation in Glyphosate-Resistant Soybean," *Applied Soil Ecology* 121 (2017): 11–19, https://doi.org/10.1016/j.apsoil.2017.09.015.
61. M. M. Newman et al., "Changes in Rhizosphere Bacterial Gene Expression Following Glyphosate Treatment," *Science of the Total Environment* 553 (2017): 32–41, https://doi.org/10.1016/j.scitotenv.2016.02.078.
62. Takayuki Tohge et al., "Shikimate and Phenylalanine Biosynthesis in the Green Lineage," *Frontiers in Plant Science* 4 (2013): 62, https://doi.org/10.3389/fpls.2013.00062.
63. G. Hrazdina, "Biosynthesis of Flavonoids," in: R. W. Hemingway and P. E. Laks (eds.), Plant Polyphenols, Basic Life Sciences series, vol. 59 (Boston:, Springer, 1992), https://doi.org/10.1007/978-1-4615-3476-1_4.
64. Shikha Tripathi et al., "Effects of Glyphosate on Vigna Radiata Var. Ml613 Assessed by Meiotic Behaviour, Total Protein and GST Activity," *International Journal of Agricultural Science and Research* 7, no. 2 (2017): 223–34.
65. Pedro Diaz Vivancos et al., "Perturbations of Amino Acid Metabolism Associated with Glyphosate-Dependent Inhibition of Shikimic Acid Metabolism Affect Cellular Redox Homeostasis and Alter the Abundance of Proteins Involved in Photosynthesis and Photorespiration," *Plant Physiology* 157 (2011): 256–68, https://doi.org/10.1104/pp.111.181024.
66. Tommi Vatanen et al., "Variation in Microbiome LPS Immunogenicity Contributes to Autoimmunity in Humans," *Cell* 165 (2016): 842–53, https://doi.org/10.1016/j.cell.2016.04.007.

67. Austin G. Davis-Richardson et al., "Bacteroides dorei Dominates Gut Microbiome Prior to Autoimmunity in Finnish Children at High Risk for Type 1 Diabetes," *Frontiers in Microbiology* 5 (2014): 678, https://doi.org/10.3389/fmicb.2014.00678.
68. Tommi Vatanen et al., "Variation in Microbiome LPS Immunogenicity Contributes to Autoimmunity in Humans," *Cell* 165 (2016): 842–53, https://doi.org/10.1016/j.cell.2016.04.007.
69. Tommi Vatanen et al., "Genomic Variation and Strain-Specific Functional Adaptation in the Human Gut Microbiome during Early Life," *Nature Microbiology* 4, no. 3 (2019): 470–79, https://doi.org/10.1038/s41564-018-0321-5.
70. J. Salonen et al., "Impact of Changed Cropping Practices on Weed Occurrence in Spring Cereals in Finland—A Comparison of Surveys in 1997–1999 and 2007–2009." *Weed Research* 53 (2012): 110–20, https://doi.org/10.1111/wre.12004.
71. Philip Case, "Putin Wants Russia to Become World Leader in Organic Food, Farmers Weekly, December 4, 2015, https://www.fwi.co.uk/international-agriculture/putin-wants-russia-become-world-leader-organic-food.

Chapter 4: Amino Acid Analogue

1. J. A. Sikorski and K. J. Gruys, "Understanding Glyphosate's Molecular Mode of Action with EPSP Synthase: Evidence Favoring an Allosteric Inhibitor Model," *Accounts of Chemical Research* 30, no. 1 (1997): 2–8, https://doi.org/10.1021/ar950122+.
2. T. Funke et al., "Molecular Basis for the Herbicide Resistance of Roundup Ready Crops," *Proceedings of the National Academy of Sciences U S A* 2103, no. 35 (2006): 13010–15, https://doi.org/10.1073/pnas.0603638103.
3. S. E. Antonarakis et al., "Disease-Causing Mutations in the Human Genome," *European Journal of Pediatrics* 159, Suppl 3 (2000): S173–8, https://doi.org/10.1007/pl00014395.
4. S. E. Antonarakis et al., "Disease-Causing Mutations in the Human Genome," *European Journal of Pediatrics* 159, Suppl 3 (2000): S175, https://doi.org/10.1007/pl00014395.
5. A. Moghal et al., "Mistranslation of the Genetic Code," *FEBS Letters* 588 (2014): 4305–10, https://doi.org/10.1016/j.febslet.2014.08.035; P. Schimmel, "Mistranslation and Its Control by tRNA Synthetases," *Philosophical Transactions of the Royal Society of London Series B Biological Sciences* 366 (2011): 29652971, https://doi.org/10.1098/rstb.2011.0158.
6. E. Rubenstein, "Misincorporation of the Proline Analog Azetidine-2-Carboxylic Acid in the Pathogenesis of Multiple Sclerosis: A Hypothesis," *Journal of Neuropathology & Experimental Neurology* 67, no. 11 (2008): 1035–40, https://doi.org/10.1097/NEN.0b013e31818add4a.

7. Raymond A. Sobel, "A Novel Unifying Hypothesis of Multiple Sclerosis," *Journal of Neuropathology & Experimental Neurology* 67, no 11 (2008) 1032–34, https://doi.org/10.1097/NEN.0b13e31818becal.
8. C. Bertin et al., "Grass Roots Chemistry: Meta-Tyrosine, an Herbicidal Nonprotein Amino Acid," *Proceedings of the National Academy of Sciences U S A* 104, No. 43 (2007): 16964–69, https://doi.org/10.1073/pnas.0707198104.
9. A. Herzine et al., "Perinatal Exposure to Glufosinate Ammonium Herbicide Impairs Neurogenesis and Neuroblast Migration through Cytoskeleton Destabilization," *Frontiers in Cellular Neuroscience* 10 (2016): 191, https://doi.org/10.3389/fncel.2016.00191.
10. B. J. Main et al., "The Use of L-Serine to Prevent β-Methylamino-L-Alanine (BMAA)-Induced Proteotoxic Stress *in Vitro*," Toxicon 109 (2016): 7–12, https://doi.org/10.1016/j.toxicon.2015.11.003; R. A. Dunlop et al., "The Non-Protein Amino Acid BMAA is Misincorporated into Human Proteins in Place of L-Serine Causing Protein Misfolding and Aggregation," *PLoS ONE* 8 (2013): e75376, https://doi.org/10.1371/journal.pone.0075376.
11. J. Krakauer et al., "Presence of L-Canavanine in *Hedysarum alpinum* Seeds and Its Potential Role in the Death of Chris McCandless," *Wilderness & Environmental Medicine* 26 (2015): 36–42, https://doi.org/10.1016/j.wem.2014.08.014.
12. H. Jakubowski, "Homocysteine Thiolactone: Metabolic Origin and Protein Homocysteinylation in Humans," *Journal of Nutrition* 130 (2000): 377S–381S, https://doi.org/10.1093/jn/130.2.377S.
13. An alternative postulate is chelation of manganese, the co-factor for flavin mononucleotide (FMN) reductase, which causes a deficiency in reduced FMN as the co-factor for EPSPS; another alternative is chelation of cobalt that serves as the co-factor for the enzyme that catalyzes the initial step of the shikimate pathway, or both.
14. S. Eschenburg et al., "How the Mutation Glycine96 to Alanine Confers Glyphosate Insensitivity to 5-Enolpyruvyl Shikimate-3-Phosphate Synthase from *Escherichia coli*," *Planta* 216, no. 1 (2002): 129–35, https://doi.org/10.1007/s00425-002-0908-0.
15. S. R. Padgette et al., "Site-Directed Mutagenesis of a Conserved Region of the 5-Enolpyruvylshikimate-3-Phosphate Synthase Active Site," *Journal of Biological Chemistry* 266, no. 33 (1991): 22364–69.
16. S. R. Padgette et al., "Site-Directed Mutagenesis," 22364–69.
17. T. Funke et al., "Structural Basis of Glyphosate Resistance Resulting from the Double Mutation $Thr^{97} \rightarrow Ile$ and $Pro^{101} \rightarrow Ser$ in 5-Enolpyruvylshikimate-3-Phosphate Synthase from *Escherichia coli*," *Journal of Biological Chemistry* 284, no. 15 (2009): 9854–60, https://doi.org/10.1074/jbc.M809771200.
18. H. C. Steinrücken and N. Amrhein, "5-Enolpyruvylshikimate-3-Phosphate Synthase of *Klebsiella pneumoniae* 2. Inhibition by Glyphosate

[N-(Phosphonomethyl)Glycine]," *European Journal of Biochemistry* 143 (1984): 351–57, https://doi.org/10.1111/j.1432-1033.1984.tb08379.x.

19. Fatma Betül Ayanoğlu et al., "Bioethical Issues in Genome Editing by CRISPR-Cas9 Technology," *Turkish Journal of Biology* 44, no. 2 (2020) 110–20, https://doi.org/10.3906/biy-1912-52.
20. Y. Dong et al., "Desensitizing Plant EPSP Synthase to Glyphosate: Optimized Global Sequence Context Accommodates a Glycine-to-Alanine Change in the Active Site," *Journal of Biological Chemistry* 294, no. 2 (2019): 716–25, https://doi.org/10.1074/jbc.RA118.006134.
21. Y. Dong et al., "Desensitizing Plant EPSP Synthase to Glyphosate: Optimized Global Sequence Context Accommodates a Glycine-to-Alanine Change in the Active Site," *Journal of Biological Chemistry* 294, no. 2 (2019): 716–25, https://doi.org/10.1074/jbc.RA118.006134.
22. Y. Dong et al., "Desensitizing Plant EPSP Synthase to Glyphosate: Optimized Global Sequence Context Accommodates a Glycine-to-Alanine Change in the Active Site," *Journal of Biological Chemistry* 294, no. 2 (2019): 716–25, https://doi.org/10.1074/jbc.RA118.006134.
23. M. A. Priestman et al., "5-Enolpyruvylshikimate-3-Phosphate Synthase from *Staphylococcus aureus* Is Insensitive to Glyphosate," *FEBS Letters* 579, No. 3 (2005): 728–32, https://doi.org/10.1016/j.febslet.2004.12.057.
24. W. P. Ridley and K. A. Chott, "Uptake, Depuration and Bioconcentration of C-14 Glyphosate to Bluegill Sunfish (*Lepomis machrochirus*) Part II: Characterization and Quantitation of Glyphosate and Its Metabolites," Monsanto Agricultural Company (unpublished study), August 1989.
25. Anthony Samsel and Stephanie Seneff, "Glyphosate Pathways to Modern Diseases VI: Prions, Amyloidoses and Autoimmune Neurological Diseases," *Journal of Biological Physics and Chemistry* 17 (2017): 8–32, https://doi.org/10.4024/25SA16A.jbpc.17.01.
26. C. Lowrie, "The Metabolism of [C-14] N-Acetyl-Glyphosate (IN-MCX20) in Laying Hens," Charles River Laboratories Project no.210573, submitted by E. I. DuPont de Nemours and Company, DuPont Report No. Dupont-19795 (2007); C. Lowrie, "Metabolism of [14C]-N-Acetyl-Glyphosate (IN-MCX20) in the Lactating Goat," Charles River Laboratories Project no. 210583, submitted by E. I. du Pont de Nemours and Company, DuPont Report No. DuPont-19796 (2007).
27. M. M. Newman et al., "Changes in Rhizosphere Bacterial Gene Expression Following Glyphosate Treatment," *Science of the Total Environment* 553 (2016): 32–41, https://doi.org/10.1016/j.scitotenv.2016.02.078.
28. Michael N. Antoniou et al., "Glyphosate Does Not Substitute for Glycine in Proteins of Actively Dividing Mammalian Cells," *BMC Research Notes* 12 (2019): 494, https://doi.org/10.1186/s13104-019-4534-3.

29. Monika Krüger et al., "Detection of Glyphosate Residues in Animals and Humans," *Journal of Environmental and Analytical Toxicology* 4 (2014): 210; Monika Krüger et al., "Field Investigations of Glyphosate in Urine of Danish Dairy Cows," *Journal of Environmental & Analytical Toxicology* 3 (2013): 5, http://dx.doi.org/10.4172/2161-0525.1000186; A. Aris and S. Leblanc, "Maternal and Fetal Exposure to Pesticides Associated to Genetically Modified Foods in Eastern Townships of Quebec, Canada," *Reproductive Toxicology* 31, no. 4 (2011) 528–33, 10.1016/j.reprotox.2011.02.004.
30. R. Cailleau et al., "Long-Term Human Breast Carcinoma Cell Lines of Metastatic Origin: Preliminary Characterization," *In Vitro* 14, no. 11 (1978): 911–15, https://doi.org/10.1007/BF02616120.
31. Michael N. Antoniou et al., "Glyphosate Does Not Substitute for Glycine in Proteins of Actively Dividing Mammalian Cells," *BMC Research Notes* 12 (2019): 494, https://doi.org/10.1186/s13104-019-4534-3.
32. M. J. Cope et al., "Conservation within the Myosin Motor Domain: Implications for Structure and Function," *Structure* 4, no. 8 (1996) 969–87, https://doi.org/10.1016/s0969-2126(96)00103-7.
33. F. Kinose et al., "Glycine 699 Is Pivotal for the Motor Activity of Skeletal Muscle Myosin," *Journal of Cell Biology* 134, no. 4 (1996): 895–909, https://doi.org/10.1083/jcb.134.4.895.
34. A. Richards et al., "The Substitution of Glycine 661 by Arginine in Type III Collagen Produces Mutant Molecules with Different Thermal Stabilities and Causes Ehlers-Danlos Syndrome Type IV," *Journal of Medical Genetics* 30, no. 8 (1993): 690–93, https://doi.org/10.1136/jmg.30.8.690; B. Steinmann et al., *Connective Tissue and Its Heritable Disorders, Molecular Genetic and Medical Aspects*, (New York: Wiley-Liss), 1993: 351–407.
35. Hilal Maradit Kremers et al., "Prevalence of Total Hip and Knee Replacement in the United States," *The Journal of Bone and Joint Surgery (American Volume)* 97, no. 17 (2015) 1386–97, https://doi.org/10.2106/JBJS.N.01141.
36. Sanguk Kim et al., "Transmembrane Glycine Zippers: Physiological and Pathological Roles in Membrane Proteins," *Proceedings of the National Academy of Sciences U S A* 102, no. 40 (2005): 14278–83, https://doi.org/10.1073/pnas.0501234102.
37. K. A. Matthews et al., "Racial and Ethnic Estimates of Alzheimer's Disease and Related Dementias in the United States (2015–2060) in Adults Aged ≥ 65 Years," *Alzheimer's & Dementia* 15, no. 1 (2019): 17–24, https://doi.org/10.1016/j.jalz.2018.06.3063external icon.
38. Sanguk Kim et al., "Transmembrane Glycine Zippers: Physiological and Pathological Roles in Membrane Proteins," *Proceedings of the National Academy of Sciences* 192, no. 40 (2005) 14278–83, https://doi.org/10.1073/pnas.0501234102.

39. Giampaolo Merlini et al., "Amyloidosis: Pathogenesis and New Therapeutic Options," *Journal of Clinical Oncology* 29, no. 14 (2011): 1924–33, https://doi.org/10.1200/JCO.2010.32.2271.
40. E. Wertheimer et al., "Two Mutations in a Conserved Structural Motif in the Insulin Receptor Inhibit Normal Folding and Intracellular Transport of the Receptor," *Journal of Biological Chemistry* 269 (1994): 7587–92; U. M. Koivisto et al., "A Novel Cellular Phenotype for Familial Hypercholesterolemia Due to a Defect in Polarized Targeting of LDL Receptor," *Cell* 105, no. 5 (2001): 575–85, PMID: 8125981.
41. Duane Graveline, "Adverse Effects of Statin Drugs: A Physician Patient's Perspective," *Journal of American Physicians and Surgeons* 20, no. 1 (2015): 7–11.

Chapter 5: The Phosphate Puzzle

1. Nuria de María et al., "New Insights on Glyphosate Mode of Action in Nodular Metabolism: Role of Shikimate Accumulation," *Journal of Agricultural and Food Chemistry* 54 (2006): 2621–28, https://doi.org/10.1021/jf058166c.
2. F. Peixoto, "Comparative Effects of the Roundup and Glyphosate on Mitochondrial Oxidative Phosphorylation," *Chemosphere* 61 (2004) 1115–22, https://doi.org/10.1016/j.chemosphere.2005.03.044; Daiane Cattani et al., "Mechanisms Underlying the Neurotoxicity Induced by Glyphosate-Based Herbicide in Immature Rat Hippocampus: Involvement of Glutamate Excitotoxicity," *Toxicology* 329 (2014): 34–45, https://doi.org/10.1016/j.tox.2014.03.001.
3. Jagat S. Chauhan et al., "Identification of ATP Binding Residues of a Protein from its Primary Sequence," *BMC Bioinformatics* 10 (2009): 434, https://doi.org/10.1186/1471-2105-10-434.
4. Quang Khai Huynh et al., "Mechanism of Inactivation of Escherichia coli 5-Enolpyruvoylshikimate-3-Phosphate Synthase by o-Phthalaldehyde," *Journal of Biological Chemistry* 265, no. 12 (1990): 6700–4.
5. Sampath Koppole et al., "The Structural Coupling between ATPase Activation and Recovery Stroke in the Myosin II Motor," *Structure* 15 (2007): 825–37, https://doi.org/10.1016/j.str.2007.06.008.
6. T. Kambara and T. E. Rhodes, "Functional Significance of the Conserved Residues in the Flexible Hinge Region of the Myosin Motor Domain," *Journal of Biological Chemistry* 274, no. 23 (1999): 16400–6, https://doi.org/10.1074/jbc.274.23.16400.
7. Wei Lu et al., "Genome-Wide Transcriptional Responses of *Escherichia coli* to Glyphosate, a Potent Inhibitor of the Shikimate Pathway Enzyme 5-Enolpyruvylshikimate-3-Phosphate Synthase," *Molecular BioSystems* 9 (2013): 522–30, https://doi.org/10.1128/JB.01990-07.

8. Christopher M. Smith et al., "The Catalytic Subunit of cAMP-Dependent Protein Kinase: Prototype for an Extended Network of Communication," *Progress in Biophysics & Molecular Biology* 71 (1999): 313–41.
9. B. D. Grant et al., "Kinetic Analyses of Mutations in the Glycine-Rich Loop of cAMP- Dependent Protein Kinase," *Biochemistry* 37 (1998): 7708–15, https://doi.org/10.1021/bi972987w; Helen Davies et al., "Mutations of the BRAF Gene in Human Cancer," *Nature* 417, no. 6892 (2002) 949–54, https://doi.org/10.1038/nature00766.
10. Christopher M. Smith et al., "The Catalytic Subunit of cAMP-Dependent Protein Kinase: Prototype for an Extended Network of Communication," *Progress in Biophysics & Molecular Biology* 71 (1999): 313–41; Helen Davies et al., "Mutations of the BRAF Gene in Human Cancer," *Nature* 417, no. 6892 (2002) 949–54, https://doi.org/10.1038/nature00766; Wolfram Hemmer et al., "Role of the Glycine Triad in the ATP-binding Site of cAMP-dependent Protein Kinase," *Journal of Biological Chemistry* 272, no. 27 (1997): 16946–54, https://doi.org/ 10.1038/nature00766; D. Chaillot et al., "Mutation of Recombinant Catalytic Subunit of the Protein Kinase CK2 that Affects Catalytic Efficiency and Specificity," *Protein Engineering* 13 (2000): 291–98, https://doi.org/10.1093/protein/13.4.291.
11. D. Chaillot et al., "Mutation of Recombinant," 291–98.
12. V. Singh et al., "Phosphorylation: Implications in Cancer," *Protein Journal* 36 no. 1 (2017): 1–6; J. Z. Wang et al., "Abnormal Hyperphosphorylation of Tau: Sites, Regulation, and Molecular Mechanism of Neurofibrillary Degeneration," *Journal of Alzheimer's Disease* 33 (2013): S123–39.
13. Helen Davies et al., "Mutations of the BRAF Gene in Human Cancer," *Nature* 417, no. 6892 (2002): 949–54.
14. Patrick G. Gallagher et al., "Diagnosis of Pyruvate Kinase Deficiency," *Pediatric Blood & Cancer* 63, no. 5 (2016): 771–72, https://doi.org/10.1038/nature00766.
15. Unlike red blood cells, most living cells get most of their energy supply through oxidative phosphorylation of glucose and other nutrients in the mitochondria.
16. Anna Demina et al., "Six Previously Undescribed Pyruvate Kinase Mutations Causing Enzyme Deficiency," *Blood* 92, no. 2 (1998): 647–52, https://doi.org/10.1182/blood.V92.2.647.
17. John Knight et al., "Ascorbic Acid Intake and Oxalate Synthesis," *Urolithiasis* 44, no. 4 (2016): 289–97, https://doi.org/10.1007/s00240-016-0868-7.
18. S. K. Jain, "Glutathione and Glucose-6-Phosphate Dehydrogenase Deficiency can Increase Protein Glycosylation," *Free Radial Biology and Medicine* 24, no. 1 (1998): 197–201, https://doi.org/10.1016/s0891-5849(97)00223-2.
19. M. Ahdab-Barmada and J. Moossy, "The Neuropathology of Kernicterus in the Premature Neonate: Diagnostic Problems," *Journal of Neuropathology &*

Experimental Neurology 43, no. 1 (1984): 45–56, https://doi.org/10.1097/00005072-198401000-00004.

20. Sanjiv B. Amin et al., "Is Neonatal Jaundice Associated with Autism Spectrum Disorders: A Systematic Review," *Journal of Autism and Developmental Disorders* 41, no. 11 (2011): 1455–1463, https://doi.org/10.1007/s10803-010-1169-6; S. M. Al-Salehi and M. Ghaziuddin, "G6PD Deficiency in Autism: A Case-Series from Saudi Arabia," European Child & Adolescent Psychiatry 18, no. 4 (2009): 227-30, https://doi.org/0.1007/s00787-008-0721-9.
21. Anna D. Cunningham et al., "Coupling between Protein Stability and Catalytic Activity Determines Pathogenicity of G6PD Variants," *Cell Reports* 18 (2017): 2592–2599, https://doi.org/10.1016/j.celrep.2017.02.048.
22. Daiane Cattani et al., "Mechanisms Underlying the Neurotoxicity Induced by Glyphosate-Based Herbicide in Immature Rat Hippocampus: Involvement of Glutamate Excitotoxicity," *Toxicology* 329 (2014): 34–45, https://doi.org/10.1016/j.tox.2014.03.001.
23. M. Kotaka et al., "Structural Studies of Glucose-6-Phosphate and NADP+ Binding to Human Glucose-6-Phosphate Dehydrogenase," *Acta Crystallographica D* 61 (2005): 495–504, https://doi.org/10.1107/S0907444905002350.
24. X. T. Wang et al., "What Is the Role of the Second 'Structural' NADP+-Binding Site in Human Glucose-6-Phosphate Dehydrogenase?" *Protein Science* 17, no. 8 (2008): 1403–11, https://doi.org/10.1110/ps.035352.108.
25. Omar M. E. Abdel-Salam et al., "Nuclear Factor-Kappa B and Other Oxidative Stress Biomarkers in Serum of Autistic Children," *Open Journal of Molecular and Integrative Physiology* 5 (2015): 18–27, https://doi.org/10.4236/ojmip.2015.51002.
26. Lynn M. Kitchen et al., "Inhibition of δ-Aminolevulinic Acid Synthesis by Glyphosate," *Weed Science* 29, no. 5 (1981): 571–77, https://doi.org/10.1017/S004317450006375X.
27. Henry N. Kirkman et al., "Mechanisms of Protection of Catalase by NADPH: Kinetics and Stoichiometry," *The Journal of Biological Chemistry* 274 (1999): 13908–14, https://doi.org/10.1074/jbc.274.20.13908.
28. A. Ghanizadeh et al., "Glutathione-Related Factors and Oxidative Stress in Autism, a Review," *Current Medicinal Chemistry* 19, no. 23 (2012): 4000–5, https://doi.org/10.2174/092986712802002572.
29. Audrey Gehin et al., "Glyphosate-Induced Antioxidant Imbalance in HaCaT: The Protective Effect of Vitamins C and E," *Environmental Toxicology & Pharmacology* 22, no. 1 (2006): 27–34, https://doi.org/10.1016/j.etap.2005.11.003.
30. R. Turkmen et al., "Protective Effects of Resveratrol on Biomarkers of Oxidative Stress, Biochemical and Histopathological Changes Induced by Sub-Chronic Oral Glyphosate-Based Herbicide in Rats," *Toxicology Research (Cambridge)* 8, no. 2 (2019): 238–45, https://doi.org/10.1039/c8tx00287h.

31. D. Vivancos et al., "Perturbations of Amino Acid Metabolism Associated with Glyphosate-Dependent Inhibition of Shikimic Acid Metabolism Affect Cellular Redox Homeostasis and Alter the Abundance of Proteins Involved in Photosynthesis and Photorespiration," *Plant Physiology* 157, no. 1 (2011): 256–68, https://doi.org/10.1104/pp.111.181024.
32. These are G149, G196, G255, and G387.
33. S. Jill James et al., "Metabolic Biomarkers of Increased Oxidative Stress and Impaired Methylation Capacity in Children with Autism," *American Journal of Clinical Nutrition* 80 (2004): 1611–17, https://doi.org/10.1093/ajcn/80.6.1611.
34. Mostafa I. Waly, "Redox-Methylation Theory and Autism," In *The Comprehensive Guide to Autism,* Patel et al. (eds.) (New York: Springer, January 2014).
35. Eicosanoids (prostaglandins, thromboxanes, leukotrienes, endocannabinoids).
36. Ningwu Huang et al., "Diversity and Function of Mutations in P450 Oxidoreductase in Patients with Antley-Bixler Syndrome and Disordered Steroidogenesis," *American Journal of Human Genetics* 76 (2005): 729–49, https://doi.org/10.1086/429417.
37. David C. Lamb et al., "A Second FMN Binding Site in Yeast NADPH-Cytochrome P450 Reductase Suggests a Mechanism of Electron Transfer by Diflavin Reductases," *Structure* 14 (2006): 51–61, https://doi.org/10.1016/j.str.2005.09.015.
38. Lin Wu et al., "Conditional Knockout of the Mouse NADPH-Cytochrome P450 Reductase Gene," *Genesis* 36, no. 4 (2003): 177–81, https://doi.org/10.1002/gene.10214.
39. Robin Mesnage et al., "Multiomics Reveal Non-Alcoholic Fatty Liver Disease in Rats Following Chronic Exposure to an Ultra-Low Dose of Roundup Herbicide," *Scientific Reports* 7 (2017): 39328, https://doi.org/10.1038/srep39328.
40. David C. Lamb et al., "A Second FMN Binding Site in Yeast NADPH-Cytochrome P450 Reductase Suggests a Mechanism of Electron Transfer by Diflavin Reductases," *Structure* 14 (2006): 51–61, https://doi.org/10.1016/j.str.2005.09.015.
41. Q. Zhao et al., "Crystal Structure of the FMN-Binding Domain of Human Cytochrome P450 Reductase at 1.93 A Resolution," *Protein Science* 8, no. 2 (1999): 298–306, https://doi.org/10.1110/ps.8.2.298.
42. Eino Hietanen et al., "Effects of Phenoxyherbicides and Glyphosate on the Hepatic and Intestinal Biotransformation Activities in the Rat," *Acta Pharmacologica et Toxicologica* 53 (1983): 103–12, https://doi.org/10.1111/j.1600-0773.1983.tb01876.x.
43. A. W. Segal, "The NADPH Oxidase and Chronic Granulomatous Disease," *Molecular Medicine Today* 2, no. 3 (1996): 129–35.
44. Franck Debeurme et al., "Regulation of NADPH Oxidase Activity in Phagocytes: Relationship between FAD/NADPH Binding and Oxidase Complex

Assembly," *Journal of Biological Chemistry* 285, no. 43 (2010): 33197–208, https://doi.org/10.1016/1357-4310(96)88723-5, https://doi.org/10.1074/jbc.M110.151555.
45. L. M. Siegel et al., "*Escherichia coli* Sulfite Red=uctase Hemoprotein Subunit. Prosthetic Groups, Catalytic Parameters, and Ligand Complexes," *Journal of Biological Chemistry* 257, no. 11 (1982): 6343–50.
46. Julien Loubinoux et al., "Sulfate-Reducing Bacteria in Human Feces and Their Association with Inflammatory Bowel Diseases," *EMS Microbiology Ecology* 40, no. 2 (2002): 107–12, https://doi.org/10.1111/j.1574-6941.2002.tb00942.x; Sydney M. Finegold, "Desulfovibrio Species Are Potentially Important in Regressive Autism," *Medical Hypotheses* 77, no. 2 (2011): 270–74, https://doi.org/10.1016/j.mehy.2011.04.032.
47. Michael N. Antoniou et al., "Glyphosate Does Not Substitute for Glycine in Proteins of Actively Dividing Mammalian Cells," *BMC Research Notes* 12 (2019): 494, https://doi.org/10.1186/s13104-019-4534-3.

Chapter 6: Sulfate: Miracle Worker

1. D. Wacey et al., "Microfossils of Sulphur-Metabolizing Cells in 3.4-Billion-Year-Old Rocks of Western Australia," *Nature Geoscience* 4, no. 10 (2011): 698–702, https://doi.org/10.1038/ngeo1238.
2. J. M. Olson and R. E. Blankenship, "Thinking about the Evolution of Photosynthesis," *Photosynthesis Research* 80 (2004): 373–86, https://doi.org/10.1023/B:PRES.0000030457.06495.83.
3. Nikolaos Samaras et al., "A Review of Age-Related Dehydroepiandrosterone Decline and Its Association with Well-Known Geriatric Syndromes: Is Treatment Beneficial?" *Rejuvenation Res* 16, no. 4 (2013) 285–94, https://doi.org/10.1089/rej.2013.1425.
4. Stephanie Seneff and Gregory Nigh, "Glyphosate and Anencephaly: Death by a Thousand Cuts," *Journal of Neurology and Neurobiology* 3 (2017): 2, http://dx.doi.org/10.16966/2379-7150.140.
5. Lance P. Walsh et al., "Roundup Inhibits Steroidogenesis by Disrupting Steroidogenic Acute Regulatory (StAR) Protein Expression," *Environmental Health Perspectives* 108, no. 8 (2000): 769–76, https://doi.org/10.1289/ehp.00108769.
6. E. Tierney et al., "Abnormalities of Cholesterol Metabolism in Autism Spectrum Disorders," *American Journal of Human Genetics & Neuropsychiatric Genetics* 1418, No 6 (2006): 666–68, https://doi.org/10.1002/ajmg.b.30368.
7. Rosemary H. Waring and L. V. Klovrza, "Sulphur Metabolism in Autism," *Journal of Nutritional & Environmental Medicine* 10 (2000): 25–32.
8. J. J. Cannell, "Autism and Vitamin D," *Medical Hypotheses* 70 (2008): 750–59, https://doi.org/10.1080/13590840050000861.

9. Lucia O. Sampaio et al., "Heparins and Heparan Sulfates. Structure, Distribution and Protein Interactions," in *Insights into Carbohydrate Structure and Biological Function*, ed. Hugo Verli. (Kerala, India: Transworld Research Network, 2006), http://www.umc.br/_img/_noticias/755/artigo.pdf.
10. K. Jinesh et al., "Capillary Condensation in Atomic Scale Friction: How Water Acts like a Glue," *Physical Review Letters* 96 (2006): 166103, https://doi.org/10.1103/PhysRevLett.96.166103.
11. S. Reitsma et al., "The Endothelial Glycocalyx: Composition, Functions, and Visualization," *Pflügers Archiv* 454, no. 3 (2007): 345–59, https://doi.org/10.1007/s00424-007-0212-8.
12. Abha Sharma et al., "Effect of Health-Promoting Agents on Exclusion-Zone Size," *Dose-Response* (2018): 1–8.
13. E. E. Caldwell et al., "Importance of Specific Amino Acids in Protein Binding Sites for Heparin and Heparan Sulfate," *The International Journal of Biochemistry & Cell Biology* 28, no. 2 (1996): 203–16, https://doi.org/10.1177/1559325818796937.
14. J. D. Esko, "Glycosaminoglycan-Binding Proteins," in *Essentials of Glycobiology*, eds. A. Varki et al., (Cold Spring Harbor, NY: Cold Spring Harbor Laboratory Press, 1999).
15. Rosemary S. Mummery and Christopher C. Rider, "Characterization of the Heparin-Binding Properties of IL-61," *Journal of Immunology* 165 (2000): 5671–79, https://doi.org/10.4049/jimmunol.165.10.5671.
16. R. D. Rosenberg and P. S. Damus, "The Purification and Mechanism of Action of Human Antithrombin-heparin Cofactor," *Journal of Biological Chemistry* 248 (1973) 6490–505.
17. H. Li and U. Förstermann, "Nitric Oxide in the Pathogenesis of Vascular Disease," *Journal of Pathology* 190 no. 3 (2000): 244–54, https://doi.org/10.1002/(SICI)1096-9896(200002)190:3<244::AID-PATH575>3.0.CO;2-8.
18. Stephanie Seneff et al., "A Novel Hypothesis for Atherosclerosis as a Cholesterol Sulfate Deficiency Syndrome," *Theoretical Biology and Medical Modeling* 12 (2015): 9, https://doi.org/10.1186/s12976-015-0006-1.
19. P. Kleinbongard et al., "Red Blood Cells Express a Functional Endothelial Nitric Oxide Synthase," *Blood* 107, no. 7 (2006): 2943–51, https://doi.org/10.1182/blood-2005-10-3992.
20. Daniel Ikenna Udenwobele et al., "Myristoylation: An Important Protein Modification in the Immune Response," *Frontiers in Immunology* 8 (2017): 751, https://doi.org/10.3389/fimmu.2017.00751.
21. A. Vijay et al., "Uncoupling of eNOS Causes Superoxide Anion Production and Impairs NO Signaling in the Cerebral Microvessels of HPH-1 Mice," *Journal of Neurochemistry* 122, no. 6 (2012) 1211–18, https://doi.org/10.1111/j.1471-4159.2012.07872.x.

22. Christopher L. Bianco et al., "Investigations on the Role of Hemoglobin in Sulfide Metabolism by Intact Human Red Blood Cells," *Biochemical Pharmacology* 149 (2018): 163–73, https://doi.org/10.1016/j.bcp.2018.01.045.
23. This is described in detail in two papers I published together with colleagues on this topic: Stephanie Seneff et al., "Is Endothelial Nitric Oxide Synthase a Moonlighting Protein Whose Day Job Is Cholesterol Sulfate Synthesis? Implications for Cholesterol Transport, Diabetes and Cardiovascular Disease," *Entropy* 14 (2012): 2492–530, https://doi.org/10.3390/e14122492; Stephanie Seneff et al., "A Novel Hypothesis for Atherosclerosis as a Cholesterol Sulfate Deficiency Syndrome," *Theoretical Biology and Medical Modeling* 12 (2015): 9, https://doi.org/10.1186/s12976-015-0006-1.
24. K. A. Pritchard Jr. et al., "Native Low-Density Lipoprotein Increases Endothelial Cell Nitric Oxide Synthase Generation of Superoxide Anion," *Circulation Research* 77, no. 3 (1995) 510–18, https://doi.org/10.1161/01.res.77.3.510; O. Feron et al., "Hypercholesterolemia Decreases Nitric Oxide Production by Promoting the Interaction of Caveolin and Endothelial Nitric Oxide Synthase," *Journal of Clinical Investigation* 103, no. 6 (1999): 897–905, https://doi.org/10.1172/JCI4829.
25. S. Gao and J. Liu, "Association between Circulating Oxidized Low-Density Lipoprotein and Atherosclerotic Cardiovascular Disease," *Chronic Diseases and Translational Medicine* 3, no. 2 (2017) 89–94, https://doi.org/10.1016/j.cdtm.2017.02.008.
26. M. E. Morris and G. Levy, "Serum Concentration and Renal Excretion by Normal Adults of Inorganic Sulfate after Acetaminophen, Ascorbic Acid, or Sodium Sulfate," *Clinical Pharmacology & Therapy* 33 (1983): 529–36, https://doi.org/10.1038/clpt.1983.72.
27. H. Laue et al., "Taurine Reduction in Anaerobic Respiration of Bilophila wadsworthia RZATAU," *Applied and Environmental Microbiology* 63, no. 5 (1997): 2016–21, https://doi.org/10.1128/AEM.63.5.2016-2021.1997.
28. Mary A. Hickman et al., "Effect of Processing on Fate of Dietary [^{14}C]Taurine in Cats," *Journal of Nutrition* 120, no. 9 (1990): 995–1000, https://doi.org/10.1093/jn/120.9.995.
29. W. Lu et al., "Genome-wide Transcriptional Responses of *Escherichia coli* to Glyphosate, a Potent Inhibitor of the Shikimate Pathway Enzyme 5-Enolpyruvylshikimate-3-Phosphate Synthase," *Molecular Biosystems* 9, no. 3 (2013): 522–30, https://doi.org/10.1128/JB.01990-07.

Chapter 7: Liver Disease

1. Rachel Carson, *Silent Spring* (New York: Houghton Mifflin Company, 1962) 191.
2. Nelson Fausto et al., "Liver Regeneration," *Hepatology* 2006; 43 (51): S45–S53, https://doi.org/10.1002/hep.20969.

3. Raquel Jasper et al., "Evaluation of Biochemical, Hematological and Oxidative Parameters in Mice Exposed to the Herbicide Glyphosate-Roundup®," *Interdisciplinary Toxicology* 5, no. 3 (2012): 133–140, https://doi.org/10.2478/v10102-012-0022-5; Sanam Naz et al., "Effect of Glyphosate on Hematological and Biochemical Parameters of Rabbit (Oryctolagus cuniculus)," *Pure and Applied Biology* 8, no. 1 (2019): 78–92, http://doi.org/10.19045/bspab.2018.700166; M. Verderame and R. Scudiero, "How Glyphosate Impairs Liver Condition in the Field Lizard *Podarcis siculus* (Rafinesque-Schmaltz, 1810): Histological and Molecular Evidence," *BioMed Research International* (2019): 4746283, https://doi.org/10.1155/2019/4746283.
4. Mohamed Ahmed Fathi et al., "Disruption of Cytochrome P450 Enzymes in the Liver and Small Intestine in Chicken Embryos in Ovo Exposed to Glyphosate," *Environmental Science and Pollution Research* 27, no. 14 (2020): 16865–75, https://www.doi.org/10.1007/s11356-020-08269-3; Unchisa Intayoung et al., "Effect of Occupational Exposure to Herbicides on Oxidative Stress in Sprayers," *Safety and Health at Work* October 2020 [Epub ahead of print], https://doi.org/10.1016/j.shaw.2020.09.011; Verena J. Koller et al., "Cytotoxic and DNA-damaging Properties of Glyphosate and Roundup in Human-derived Buccal Epithelial Cells," *Archives of Toxicology* 86 (2012 February): 805–13, https://www.dio.org/10.1007/s00204-012-0804-8.
5. Laura E. Armstrong and Grace L. Guo, "Understanding Environmental Contaminants' Direct Effects on Non-alcoholic Fatty Liver Disease Progression," Current Environmental Health Reports 6, no. 3 (2019): 95–104, https://www.ncbi.nlm.nih.gov/pmc/articles/PMC6698395.
6. Alessandro Mantovani and Giovanni Targher, "Type 2 Diabetes Mellitus and Risk of Hepatocellular Carcinoma: Spotlight on Nonalcoholic Fatty Liver Disease," *Annals of Translational Medicine* 5, no. 13 (2017): 270, https://doi.org/10.21037/atm.2017.04.41.
7. R. Loomba et al., "The Global NAFLD Epidemic," *Nature Reviews Gastroenterology & Hepatology* 10 (2013): 686–90; Michael Fuchs, "Managing the Silent Epidemic of Nonalcoholic Fatty Liver Disease," *Federal Practitioner* 36, no. 1 (2019): 12–13, https://doi.org/10.1038/nrgastro.2013.171.
8. Zaki A. Sherif. "The Rise in the Prevalence of Nonalcoholic Fatty Liver Disease and Hepatocellular Carcinoma. in Nonalcoholic Fatty Liver Disease—An Update," Emad Hamdy Gad, ed., Intech Open, September 11, 2019, https://www.doi.org/10.5772/intechopen.85780.
9. Chris Estes et al., "Modeling the Epidemic of Nonalcoholic Fatty Liver Disease Demonstrates an Exponential Increase in Burden of Disease," *Hepatology* 67, no. 1 (2018): 123–33, https://doi.org/10.1002/hep.29466.

10. V. G. Agopian et al., "Liver Transplantation for Nonalcoholic Steatohepatitis: The New Epidemic," *Annals of Surgery* 256, no. 4 (2012): 624–33, https://doi.org/10.1097/SLA.0b013e31826b4b7e.
11. K. Shedlock et al., "Autism Spectrum Disorders and Metabolic Complications of Obesity," *The Journal of Pediatrics* 178 (2016): 183–87, https://doi.org/10.1016/j.jpeds.2016.07.055.
12. Paul J. Mills et al., "Glyphosate Excretion is Associated with Steatohepatitis and Advanced Liver Fibrosis in Patients with Fatty Liver Disease," *Clinical Gastroenterology* and Hepatology 18, no. 3 (2020): 741–43, https://doi.org/10.1016/j.cgh.2019.03.045.
13. G.-E. Séralini et al., "Long-term Toxicity of a Roundup Herbicide and a Roundup-Tolerant Genetically Modified Maize," *Environmental Sciences Europe* 26 (2014): 14, https://doi.org/10.1016/j.fct.2012.08.005; X. Ren et al., "Effects of Chronic Glyphosate Exposure to Pregnant Mice on Hepatic Lipid Metabolism in Offspring," *Environmental Pollution* 254, Pt A (2019): Article 112906, https://doi.org/10.1016/j.envpol.2019.07.074; Breanna Ford et al., "Mapping Proteome-wide Targets of Glyphosate in Mice," *Cell Chemical Biology* 24, no. 2 (2017): 133–40, https://doi.org/10.1016/j.chembiol.2016.12.013; Robin Mesnage et al., "Transcriptome Profile Analysis Reflects Rat Liver and Kidney Damage Following Chronic Ultra-Low Dose Roundup Exposure," *Environmental Health* 14 (2015): 70, https://doi.org/10.1186/s12940-015-0056-1.
14. G.-E. Séralini et al., "Long-Term Toxicity of a Roundup Herbicide and a Roundup-Tolerant Genetically Modified Maize," *Environmental Sciences Europe* 26 (2014): 14, https://doi.org/10.1016/j.fct.2012.08.005.
15. Robin Mesnage et al., "Potential Toxic Effects of Glyphosate and Its Commercial Formulations below Regulatory Limits," *Food and Chemical Toxicology* 84 (2015): 133–53, https://doi.org/10.1016/j.fct.2015.08.012.
16. Karen Briere, "Glyphosate on Feed Affects Livestock: Vet," October 19, 2017, https://www.producer.com/livestock/glyphosate-on-feed-affects-livestock-vet/.
17. Sanam Naz et al., "Effect of Glyphosate on Hematological and Biochemical Parameters of Rabbit (Oryctolagus cuniculus)," *Pure and Applied Biology* 9, no. 1 (2019): 78–92, https://doi.org/10.2478/v10102-012-0022-5.
18. K. Çavuşoğlu et al., "Protective Effect of Ginkgo biloba L. Leaf Extract against Glyphosate Toxicity in Swiss Albino Mice," Journal of Medicinal Food 14, no. 10 (2011): 1263–72, https://doi.org/10.1089/jmf.2010.0202.
19. X. Ren et al., "Effects of Chronic Glyphosate Exposure to Pregnant Mice on Hepatic Lipid Metabolism in Offspring," *Environmental Pollution* 254, Pt A (2019): 112906, https://doi.org/10.1016/j.envpol.2019.07.074.
20. Robin Mesnage et al., "Multiomics Reveal Non-Alcoholic Fatty Liver Disease in Rats Following Chronic Exposure to an Ultra-Low Dose of Roundup

Herbicide," *Scientific Reports* 7 (2017): 39328, https://doi.org/10.1038/srep39328.

21. Shaimaa M. M. Saleh et al., "Hepato-Morpholoy and Biochemical Studies on the Liver of Albino Rats after Exposure to Glyphosate- Roundup®," *The Journal of Basic and Applied Zoology* 79 (2018): 48, https://doi.org/10.1186/s41936-018-0060-4.
22. N. Soudani et al., "Glyphosate Disrupts Redox Status and Up-Regulates Metallothionein I and II Genes Expression in the Liver of Adult Rats. Alleviation by Quercetin," *General Physiology and Biophysics* 38, no. 2 (2019): 123–34, https://doi.org/10.4149/gpb_2018043.
23. M. Uotila et al., "Induction of Glutathione S-Transferase Activity and Glutathione Level in Plants Exposed to Glyphosate," *Physiologia Plantarum* 93 (1995): 689–94, https://doi.org/10.1093/toxsci/62.1.54.
24. R. Turkmen et al., "Protective Effects of Resveratrol on Biomarkers of Oxidative Stress, Biochemical and Histopathological Changes Induced by Sub-Chronic Oral Glyphosate-Based Herbicide in Rats," *Toxicology Research (Cambridge)* 8, no. 2 (2019): 238–45, https://doi.org/10.1039/c8tx00287h.
25. Raquel Jasper et al., "Evaluation of Biochemical, Hematological and Oxidative Parameters in Mice Exposed to the Herbicide Glyphosate-Roundup," *Interdisciplinary Toxicology* 5, no. 3 (2012): 133140, https://doi.org/10.2478/v10102-012-0022-5.
26. Stephanie Seneff et al., "Can Glyphosate's Disruption of the Gut Microbiome and Induction of Sulfate Deficiency Explain the Epidemic in Gout and Associated Diseases in the Industrialized World?" *Journal of Biological Physics and Chemistry* 17 (2017): 53–76, https://www.doi.org/10.4024/04SE17A.jbpc.17.02.
27. Gerald M. Carlson and Todd Holyoak, "Structural Insights into the Mechanism of Phosphoenolpyruvate Carboxykinase Catalysis," *Journal of Biological Chemistry* 284, no. 4 (2009): 27037–41, https://doi.org/10.1074/jbc.R109.040568.
28. P. T. Clayton et al., "Mitochondrial Phosphoenolpyruvate Carboxykinase Deficiency," *European Journal of Pediatrics* 145, nos. 1–2 (1986): 46–50, https://doi.org/10.1007/BF00441851.
29. As I first explained in chapter 3.
30. Monika Krüger et al., "Field Investigations of Glyphosate in Urine of Danish Dairy Cows," *Journal of Environmental & Analytical Toxicology* 3 (2013): 5, http://dx.doi.org/10.4172/2161-0525.1000186.
31. M. M. Newman et al., "Changes in Rhizosphere Bacterial Gene Expression Following Glyphosate Treatment," *Sci Total Environment* 553 (2017): 32–41, https://doi.org/10.1016/j.scitotenv.2016.02.078.
32. D. L. Baly et al., "Pyruvate Carboxylase and Phosphoenolpyruvate Carboxykinase Activity in Developing Rats: Effect of Manganese Deficiency," *Journal of*

Nutrition 115, no. 7 (1985): 872–79, https://doi.org/10.1203/00006450-197901000-00009.

33. Parvin Hakimi et al., "Phosphoenolpyruvate Carboxykinase and the Critical Role of Cataplerosis in the Control of Hepatic Metabolism," *Nutrition and Metabolism* 2 (2005): 33, https://doi.org/10.1186/1743-7075-2-33.
34. Parvin Hakimi et al., "Phosphoenolpyruvate Carboxykinase," 33, 5.
35. Fernando Rafael de Moura et al., "Effects of Glyphosate-Based Herbicide on Pintado da Amazônia: Hematology, Histological Aspects, Metabolic Parameters and Genotoxic Potential," *Environmental Toxicology and Pharmacology* 56 (2017): 241–48, https://doi.org/10.1016/j.etap.2017.09.019.
36. Aparamita Pandey et al., "Inflammatory Effects of Subacute Exposure of Roundup in Rat Liver and Adipose Tissue," *Dose-Response* (2019): 1–11, https://doi.org/10.1177/1559325819843380.
37. Pengxiang She et al., "Phosphoenolpyruvate Carboxykinase Is Necessary for the Integration of Hepatic Energy Metabolism," *Molecular and Cellular Biology* 20, no. 17 (2000): 6508–17, https://doi.orig/10.1128/mcb.20.17.6508-6517.2000.
38. V. R. Anjali and C. Aruna Devi, "Impact of Glyphosate on Intermediary and Mitochondrial Metabolism on a Freshwater Fish, Aplocheilus lineatus (Valenciennes)," *Journal of Aquatic Biology & Fisheries* 5 (2017): 27–35.
39. Shawn C. Burgess et al., "Impaired Tricarboxylic Acid Cycle Activity in Mouse Livers Lacking Cytosolic Phosphoenolpyruvate Carboxykinase," *Journal of Biological Chemistry* 279, no. 47 (2004): 48941–49, https://doi.org/10.1074/jbc.M407120200.
40. Vidhu Gill et al., "Advanced Glycation End Products (AGEs) May Be a Striking Link between Modern Diet and Health," *Biomolecules* 9, no. 12 (2019): Article 888, https://doi.org/10.3390/biom9120888.
41. Richard W. Hanson and Parvin Hakimi, "Born to Run: The Story of the PEPCK-Cmus Mouse," *Biochimie* 90 (2008): 838–42, https://doi.org/10.1016/j.biochi.2008.03.009.
42. Rob Cook, "World Beef Production: Ranking of Countries," August 5, 2020, https://beef2live.com/story-world-beef-production-ranking-countries-0-106885; United States Department of Agriculture. "Land Use: Range & Pasture," https://www.nrcs.usda.gov/wps/portal/nrcs/main/national/landuse/rangepasture/.
43. S. E. Mills, "Biological Basis of the Ractopamine Response," *Journal of Animal Science* 80 (2002): E28–E32.
44. Georges Bories et al., "Safety Evaluation of Ractopamine: Scientific Opinion of the Panel on Additives and Products or Substances used in Animal Feed," *The European Food Safety Authority Journal* 1041 (2009): 1–52, https://doi.org/10.2527/animalsci2002.80E-Suppl_2E28x.

45. D. M. Brown et al., "Mitochondrial Phosphoenolpyruvate Carboxykinase (PEP-CK-M) and Serine Biosynthetic Pathway Genes Are Co-ordinately Increased during Anabolic Agent-Induced Skeletal Muscle Growth," *Scientific Reports* 6 (2016): 28693, https://doi.org/10.1038/srep28693.

Chapter 8: Reproduction and Early Development

1. Naina Kumar and Amit Kant Singh, "Trends of Male Factor Infertility, an Important Cause of Infertility: A Review of Literature," *Journal of Human Reproductive Sciences* 8, no. 4 (2015): 191–96, https://doi.org/10.4103/0974-1208.170370.
2. Centers for Disease Control and Prevention, "State-Specific Assisted Reproductive Technology Surveillance," https://www.cdc.gov/art/state-specific-surveillance/index.html.
3. Gaby Galvin, "New Safety Standards Aim to Improve the Quality of Hospital Maternal Care," US News, August 29, 2019, last accessed Sept. 27, 2019, https://www.usnews.com/news/health-news/articles/2019-08-29/new-safety-standards-aim-to-improve-the-quality-of-hospital-maternal-care.
4. Lovney Kanguru et al., "The Burden of Obesity in Women of Reproductive Age and in Pregnancy in a Middle-income Setting: A Population Based Study from Jamaica," *PLoS One* 12, no. 12 (2017): e0188677, https://doi.org/10.1371/journal.pone.0188677; Ana Paula Esteves-Pereira et al., "Caesarean Delivery and Postpartum Maternal Mortality: A Population-Based Case Control Study in Brazil," *PLoS One* 11, no. 4 (2016): e0153396, https://doi.org/10.1371/journal.pone.0153396; Martin A. Makary and Michael Daniel, "Medical Error—The Third Leading Cause of Death in the US," *British Medical Journal* 353 (2016): i2139, https://doi.org/10.1136/bmj.i2139; Solwayo Ngwenya, "Postpartum Hemorrhage: Incidence, Risk Factors, and Outcomes in a Low-resource Setting," *International Journal of Women's Health* 8 (2016): 647–50, https://doi.org/10.2147/IJWH.S119232.
5. Childrens' Health Defense Team, "Infant and Child Mortality in the U.S. Nothing to Brag About," September 26, 2019, last accessed September 26, 2019, https://childrenshealthdefense.org/news/infant-and-child-mortality-in-the-u-s-nothing-to-brag-about/.
6. Ronit Machtinger et al., "Bisphenol-A and Human Oocyte Maturation in Vitro," *Human Reproduction* 28, no. 10 (2013): 2735–45, https://doi.org/10.1093/humrep/det312; Margaux McBirney et al., "Atrazine Induced Epigenetic Transgenerational Inheritance of Disease, Lean Phenotype and Sperm Epimutation Pathology Biomarkers," *PLoS ONE* 12, no. 9 (2017): e0184306, https://doi.org/10.1371/journal.pone.0184306.
7. S. Parvez et al., "Glyphosate Exposure in Pregnancy and Shortened Gestational Length: A Prospective Indiana Birth Cohort Study," *Environmental Health* 17 (2018): 23, https://doi.org/10.1186/s12940-018-0367-0.

8. P. Kongtip et al., "Glyphosate and Paraquat in Maternal and Fetal Serums in Thai Women," *Journal of Agromedicine* 22 (2017): 282–89, https://doi.org/10.1080/1059924X.2017.1319315.
9. Maria M. Milesi et al., "Perinatal Exposure to a Glyphosate-Based Herbicide Impairs Female Reproductive Outcomes and Induces Second-Generation Adverse Effects in Wistar Rats," *Archives of Toxicology* 92, no. 8 (2018): 2629–43, https://doi.org/10.1007/s00204-018-2236-6.
10. Laura N. Vandenberg et al., "Hormones and Endocrine-Disrupting Chemicals: Low-Dose Effects and Nonmonotonic Dose Responses," *Endocrine Reviews* 33, no. 3 (2012): 378–455, https://doi.org/10.1210/er.2011-1050.
11. Saniya Rattan, "Exposure to Endocrine Disruptors during Adulthood: Consequences for Female Fertility," *Journal of Endocrinology* 233, no. 3 (2017): R109–R129, https://doi.org/10.1530/JOE-17-0023.
12. Eliane Dallegrave et al., "The Teratogenic Potential of the Herbicide Glyphosate—Roundup in Wistar Rats," *Toxicology Letters* 142 (2003): 45–52, https://doi.org/10.1016/s0378-4274(02)00483-6.
13. Juan P. Muñoz et al., "Glyphosate and the Key Characteristics of an Endocrine Disruptor: A Review," *Chemosphere* October 19, 2020 [Epub ahead of print], https://doi.org/10.1016/j.chemosphere.2020.128619.
14. George Anifandis et al., "The In Vitro Impact of the Herbicide Roundup on Human Sperm Motility and Sperm Mitochondria," *Toxics* 6 (2018): Article 2, https://www.doi.org/10.3390/toxics6010002.
15. Tyrone B. Hayes et al., "Atrazine Induces Complete Feminization and Chemical Castration in Male African Clawed Frogs (*Xenopus laevis*)," *Proceedings of the National Academy of Sciences of the United States of America* 107, no. 10 (2010): 4612–17, https://doi.org/10.1073/pnas.0909519107.
16. S. O. Abarikwu et al., "Combined Effects of Repeated Administration of Bretmont Wipeout (Glyphosate) and Ultrazin (Atrazine) on Testosterone, Oxidative Stress and Sperm Quality of Wistar Rats," *Toxicology Mechanisms and Methods* 25 (2015): 70–80, https://doi.org/10.3109/15376516.2014.989349.
17. Note: The lowest exposure of 0.5 milligram per kilogram of body weight per day is considered the Acceptable Daily Intake for glyphosate by some regulators. It also happens to be 3.5 times *lower* than the US Acceptable Daily Intakes (ADI) of 1.75 milligrams per kilogram of body weight per day, as established by the EPA. (You read that right: The US EPA allows for much more glyphosate contamination to be considered "safe" than our neighbors in the rest of the world.) Five milligrams per kilogram of body weight per day is an amount considered to have no observed adverse effects.

18. T. H. Pham et al., "Perinatal Exposure to Glyphosate and a Glyphosate-Based Herbicide Affect Spermatogenesis in Mice," *Toxicological Sciences* 169, no. 1 (2019): 260–71, https://doi.org/10.1093/toxsci/kfz039.
19. Fabiana Manservisi et al., "The Ramazzini Institute 13-Week Pilot Study Glyphosate-Based Herbicides Administered at Human-Equivalent Dose to Sprague Dawley Rats: Effects on Development and Endocrine System," *Environmental Health* 18 (2019): 15, https://doi.org/10.1186/s12940-019-0453-y.
20. Jakeline Liara Teleken et al., "Glyphosate-Based Herbicide Exposure during Pregnancy and Lactation Malprograms the Male Reproductive Morphofunction in F1 Offspring," *Journal of Developmental Origins of Health and Disease* 16 (2019): 1–8, https://doi.org/10.1017/S2040174419000382.
21. Sustainable Health, "New Studies Show Glyphosate Causes Reproductive Health Damage," July 13, 2020, https://sustainablepulse.com/2020/07/13/new-studies-show-glyphosate-causes-infertility-and-reproductive-health-damage/.
22. Lance P. Walsh et al., "Roundup Inhibits Steroidogenesis by Disrupting Steroidogenic Acute Regulatory (StAR) Protein Expression," *Environmental Health Perspectives* 108, no. 8 (2000): 769–76, https://doi.org/10.1289/ehp.00108769.
23. S. W. Ahn et al., "Phosphoenolpyruvate Carboxykinase and Glucose-6-Phosphatase Are Required for Steroidogenesis in Testicular Leydig Cells," *Journal of Biological Chemistry* 287 (2012): 41875–87, https://doi.org/10.1074/jbc.M112.421552.
24. Andrew S. Midzak et al., "ATP Synthesis, Mitochondrial Function, and Steroid Biosynthesis in Rodent Primary and Tumor Leydig Cells," *Biology of Reproduction* 84, no. 5 (2011): 976–85, https://doi.org/10.1095/biolreprod.110.087460.
25. S. W. Ahn et al., "Phosphoenolpyruvate Carboxykinase and Glucose-6-Phosphatase Are Required for Steroidogenesis in Testicular Leydig Cells," *Journal of Biological Chemistry* 287 (2012): 41875–87, https://doi.org/10.1074/jbc.M112.421552.
26. Y. H. Kim et al., "Heart Rate-Corrected QT Interval Predicts Mortality in Glyphosate-Surfactant Herbicide-Poisoned Patients," *The American Journal of Emergency Medicine* 32, no. 3 (2014): 203–7, https://doi.org/10.1016/j.ajem.2013.09.025.
27. Joe-Elie Șalem et al., "Androgenic Effects on Ventricular Repolarization: A Translational Study from the International Pharmacovigilance Database to iPSC-Cardiomyocytes," *Circulation* 140, no. 13 (2019): 1070–80, https://doi.org/10.1161/CIRCULATIONAHA.119.040162.
28. J. B. Schwartz et al., "Effects of Testosterone on the Q-T Interval in Older Men and Older Women with Chronic Heart Failure," *International Journal of Andrology* 34, no. 5 (2011): e415–21, https://doi.org/10.1111/j.1365-2605.2011.01163.x.
29. Carrie C. Dennett and Judy Simon, "The Role of Polycystic Ovary Syndrome in Reproductive and Metabolic Health: Overview and Approaches for Treatment,"

Diabetes Spectrum 28, no. 2 (2015): 116–20, https://doi.org/10.2337/diaspect.28.2.116.
30. S. M. Sirmans and K. A. Pate, "Epidemiology, Diagnosis, and Management of Polycystic Ovary Syndrome," *Clinical Epidemiology* 6 (2014): 1–13, https://doi.org/10.2147/CLEP.S37559.
31. Jie Chen et al., "The Correlation of Aromatase Activity and Obesity in Women with or without Polycystic Ovary Syndrome," *Journal of Ovarian Research* 8 (2015): 11, https://doi.org/10.1186/s13048-015-0139-1.
32. Jim Parker, "A New Hypothesis for the Mechanism of Glyphosate Induced Intestinal Permeability in the Pathogenesis of Polycystic Ovary Syndrome," *Australasian College of Nutritional & Environmental Medicine Journal* 34, no. 2 (2015): 3–7.
33. Christopher Hakim et al., "Gestational Hyperandrogenism in Developmental Programming," *Endocrinology* 158, no. 2 (2017): 199–212, https://doi.org/10.1210/en.2016-1801.
34. Adriana Cherskov et al., "Polycystic Ovary Syndrome and Autism: A Test of the Prenatal Sex Steroid Theory," *Translational Psychiatry* 8 (2018): 136, https://doi.org/10.1038/s41398-018-0186-7.
35. Kenan Qin and Robert L. Rosenfield, "Mutations of the Hexose-6-Phosphate Dehydrogenase Gene Rarely Cause Hyperandrogenemic Polycystic Ovary Syndrome," *Steroids* 76, nos. 1–2 (2011): 135–39, https://doi.org/10.1016/j.steroids.2010.10.001.
36. Jie Chen et al., "The Correlation of Aromatase Activity and Obesity in Women with or without Polycystic Ovary Syndrome," *Journal of Ovarian Research* 8 (2015): 11, https://doi.org/10.1186/s13048-015-0139-1.
37. Wilma Oostdijk et al., "PAPSS2 Deficiency Causes Androgen Excess via Impaired DHEA Sulfation: *In Vitro* and *in Vivo* Studies in a Family Harboring Two Novel PAPSS2 Mutations," *The Journal of Clinical Endocrinology & Metabolism* 100 (2015): E672–E680, https://doi.org/10.1210/jc.2014-3556.
38. Sophie Richard et al., "Differential Effects of Glyphosate and Roundup on Human Placental Cells and Aromatase," *Environmental Health Perspectives* 113, no. 6 (2005): 716–20, https://doi.org/10.1289/ehp.7728.
39. Eric L. Ding et al., "Sex Hormone-Binding Globulin and Risk of Type 2 Diabetes in Women and Men," *New England Journal of Medicine* 361, no. 12 (2009): 1152–63, https://doi.org/10.1056/NEJMoa0804381.
40. Stephanie Seneff and Gregory Nigh, "Glyphosate and Anencephaly: Death by a Thousand Cuts," *Journal of Neurology and Neurobiology* 3 (2017): 2, http://dx.doi.org/10.16966/2379-7150.140.
41. D. M. Juriloff and M. J. Harris, "Mouse Models for Neural Tube Closure Defects," *Human Molecular Genetics* 9 (2000): 993–1000, https://doi.org/10.1093/hmg/9.6.993.

42. NBC News, "'Bizarre' Cluster of Severe Birth Defects Haunts Health Experts," February 16, 2014, https://www.nbcnews.com/health/kids-health/bizarre-cluster-severe-birth-defects-haunts-health-experts-n24986.
43. Dr. Don Huber, personal communication; Barbara H. Petersen, "Glyphosate, Brain Damaged Babies, and Yakima Valley—A River Runs Through It," July 28, 2013.
44. CYP enzymes involved in metabolizing vitamin A include CYP26A1, CYP26B1, and CYP26C1.
45. A. Paganelli et al., "Glyphosate-Based Herbicides Produce Teratogenic Effects on Vertebrates by Impairing Retinoic Acid Signaling," *Chemical Research in Toxicology* 23, no. 10 (2010): 1586–95, https://doi.org/doi: 10.1021/tx1001749.
46. H. Campaña et al., "Births Prevalence of 27 Selected Congenital Anomalies in 7 Geographic Regions of Argentina" [Article in Spanish], *Archivos Argentinos de Pediatría* 108, no. 5 (2010): 409–17, https://pubmed.ncbi.nlm.nih.gov/21132229/.
47. J. Santos-Guzmán et al., "Antagonism of Hypervitaminosis A-Induced Anterior Neural Tube Closure Defects with a Methyl-Donor Deficiency in Murine Whole-Embryo Culture," *Journal of Nutrition* 133, no. 11 (2003): 3561–70, https://doi.org/10.1093/jn/133.11.3561.
48. Robin Mesnage et al., "Glyphosate Exposure in a Farmer's Family," *Journal of Environmental Protection* 3 (2012): 1001–3, http://dx.doi.org/10.4236/jep.2012.39115.
49. Deepika Kubsad et al., "Assessment of Glyphosate Induced Epigenetic Transgenerational Inheritance of Pathologies and Sperm Epimutations: Generational Toxicology," *Scientific Reports* 9 (2019): 6372, https://doi.org/10.1038/s41598-019-42860-0.
50. Deepika Kubsad et al., "Assessment of Glyphosate," 6372.
51. Maria M. Milesi et al., "Perinatal Exposure to a Glyphosate-Based Herbicide Impairs Female Reproductive Outcomes and Induces Second-Generation Adverse Effects in Wistar Rats," *Archives of Toxicology* 92, no. 8 (2018): 2629–43, https://doi.org/10.1007/s00204-018-2236-6.
52. Oren R. Lyons Jr., Native American Faithkeeper of the Turtle Clan of the Seneca Nations of the Iroquois Confederacy. See: https://motivateus.com/cibt.htm.
53. M. Antoniou et al., "Teratogenic Effects of Glyphosate-Based Herbicides: Divergence of Regulatory Decisions from Scientific Evidence," *Journal of Environmental & Analytical Toxicology* S4 (2012): 006; Dianne Cattani et al., "Mechanisms Underlying the Neurotoxicity Induced by Glyphosate-Based Herbicide in Immature Rat Hippocampus: Involvement of Glutamate Excitotoxicity," *Toxicology* 320 (2014): 34–45, https://doi.org/10.1016/j.tox.2014.03.001; J. S. de Souza et al., "Perinatal Exposure to Glyphosate-Based Herbicide Alters the Thyrotrophic Axis and Causes Thyroid Hormone Homeostasis Imbalance in Male Rats," *Toxicology* 377 (2017): 25–37, https://doi.org/10.1016/j.tox.2016.11.005; X. Ren et al., "Effects of Chronic Glyphosate Exposure to Pregnant Mice on Hepatic

Lipid Metabolism in Offspring," *Environmental Pollution* 254, Pt A (2019): 112906, https://doi.org/10.1016/j.envpol.2019.07.074.

Chapter 9: Neurological Disorders

1. Vincent Planche et al., "Acute Toxic Limbic Encephalopathy Following Glyphosate Intoxication," *Neurology* 92, no. 1 (2019): 533–36, https://doi.org/10.1212/WNL.0000000000007115.
2. Xianlin Han, "Potential Mechanisms Contributing to Sulfatide Depletion at the Earliest Clinically Recognizable Stage of Alzheimer's Disease: A Tale of Shotgun Lipidomics," *Journal of Neurochemistry* 103, Suppl 1 (2007): 171–79, https://doi.org/10.1111/j.1471-4159.2007.04708.x.
3. Ellen A. Kramarow and Betzaida Tejada-Vera, "Dementia Mortality in the United States, 2000–2017," *National Vital Statistics Report* 68, no. 2 (2019): 1–29.
4. Ellen A. Kramarow and Betzaida Tejada-Vera, "Dementia Mortality," 1–29.
5. Joseph Guan and Guohua Li, "Injury Mortality in Individuals with Autism," American Journal of Public Health, April 2017.
6. L. P. Heilbrun et al., "Maternal Chemical and Drug Intolerances: Potential Risk Factors for Autism and Attention Deficit Hyperactivity Disorder (ADHD)," *Journal of the American Board of Family Medicine* 23, no. 4 (2015): 461–70, https://doi.org/10.3122/jabfm.2015.04.140192.
7. Sabrina Rossi and Alessio Pitidis, "Multiple Chemical Sensitivity Review of the State of the Art in Epidemiology, Diagnosis, and Future Perspectives," *Journal of Occupational and Environmental Medicine* 60, no. 2 (2018): 138–46, https://doi.org/10.1097/JOM.0000000000001215.
8. Olav Albert Christophersen, "Should Autism Be Considered a Canary Bird Telling That *Homo sapiens* May Be on Its Way to Extinction?" *Microbial Ecology in Health & Disease* 23 (2012): 19008, https://www.ncbi.nlm.nih.gov/pmc/articles/PMC3747741/.
9. Cynthia D. Nevison, "A Comparison of Temporal Trends in United States Autism Prevalence to Trends in Suspected Environmental Factors," *Environmental Health* 13 (2014): 73, https://doi.org/10.1186/1476-069X-13-73.
10. Romain Kroum Gherardi et al., "Biopersistence and Brain Translocation of Aluminum Adjuvants of Vaccines," *Frontiers of Neurology* 6 (2015): 4, https://doi.org/10.3389/fneur.2015.00004.
11. Lucio G. Costa and Gennaro Giordano, "Developmental Neurotoxicity of Polybrominated Diphenyl Ether (PBDE) Flame Retardants," *Neurotoxicology* 28, no. 6 (2007): 1047–67, https://www.doi.org/10.1016/j.neuro.2007.08.007; Julie B. Herbstman and Jennifer K. Mall, "Developmental Exposure to Polybrominated Diphenyl Ethers and Neurodevelopment," *Current Environmental Health Reports* 1, no. 2 (2014): 101–12, https://doi.org/10.1007/s40572-014-0010-3.

12. S. Seneff et al., "Aluminum and Glyphosate Can Synergistically Induce Pineal Gland Pathology: Connection to Gut Dysbiosis and Neurological Disease," *Agricultural Sciences* 6 (2015): 42–70, https://doi.org/10.4236/as.2015.61005.
13. Anna Strunecka et al., "Immunoexcitotoxicity as the Central Mechanism of Etiopathology and Treatment of Autism Spectrum Disorders: A Possible Role of Fluoride and Aluminum," *Surgical Neurology International* 9 (2018 April): article 74, https://doi.org/10.4103/sni.sni_407_17.
14. L. Tomljenovic, "Aluminum and Alzheimer's Disease: After a Century of Controversy, Is There a Plausible Link?" Journal of Alzheimer's Disease 23, no. 4 (2011): 567–98, https://doi.org/10.3233/JAD-2010-101494.
15. A. C. Alfrey et al., "Dialysis Encephalopathy Syndrome—Possible Aluminium Intoxication," *New England Journal of Medicine* 294 (1976): 184–88, https://doi.org/10.1056/NEJM197601222940402.
16. Matthew Mold et al., "Aluminium in Brain Tissue in Autism," *Journal of Trace Elements in Medicine and Biology* 46 (2018): 76–82, https://doi.org/10.1016/j.jtemb.2017.11.012.
17. Matthew Mold et al., "Intracellular Aluminium in Inflammatory and Glial Cells in Cerebral Amyloid Angiopathy: A Case Report," *International Journal of Environmental Research and Public Health* 16 (2019): 1459, https://doi.org/10.3390/ijerph16081459.
18. S. B. Lang et al., "Piezoelectricity in the Human Pineal Gland," *Bioelectrochemistry and Bioengineering* 41 (1996): 1915, https://doi.org/10.1016/S0302-4598(96)05147-1.
19. Stephanie Seneff et al., "Aluminum and Glyphosate Can Synergistically Induce Pineal Gland Pathology: Connection to Gut Dysbiosis and Neurological Disease," *Agricultural Sciences* 6 (2015): 42–70, https://doi.org/10.4236/as.2015.61005.
20. Pauline Chaste and Marion Leboyer, "Autism Risk Factors: Genes, Environment, and Gene-Environment Interactions," *Dialogues in Clinical Neuroscience* 14, no. 3 (2012): 281–92, https://doi.org/i:10.31887/DCNS.2012.14.3/pchaste.
21. Leo Kanner, "Autistic Disturbances of Affective Contact," Nervous Child 2 (1943): 217–50.
22. James Lyons-Weiler, *The Environmental and Genetic Causes of Autism* (New York: Skyhorse Publishers, 2016).
23. Ondine S. von Ehrenstein et al., "Prenatal and Infant Exposure to Ambient Pesticides and Autism Spectrum Disorder in Children: Population Based Case-Control Study," *British Medical Journal* 364 (2019): l962, https://doi.org//10.1136/bmj.l962.
24. Jiaqiang Xu et al., "The Role of L-Type Amino Acid Transporters in the Uptake of Glyphosate across Mammalian Epithelial Tissues," *Chemosphere* 145 (2016): 487–94, https://doi.org/10.1016/j.chemosphere.2015.11.062.

25. Qian Wang and Jeff Holst, "L-type Amino Acid Transport and Cancer: Targeting the mTORC1 Pathway to Inhibit Neoplasia," *American Journal of Cancer Research* 5, No 4 (2015): 1281–94.
26. Daiane Cattani et al., "Mechanisms Underlying the Neurotoxicity Induced by Glyphosate-Based Herbicide in Immature Rat Hippocampus: Involvement of Glutamate Excitotoxicity," *Toxicology* 320 (2014): 34–45, https://doi.org/10.1016/j.tox.2014.03.001.
27. Lola Rueda-Ruzafa et al., "Gut Microbiota and Neurological Effects of Glyphosate," *Neurotoxicology* 75 (2019): 1–8, https://doi.org/10.1016/j.neuro.2019.08.006.
28. Muhammad Irfan Masood et al., "Environment Permissible Concentrations of Glyphosate in Drinking Water Can Influence the Fate of Neural Stem Cells from the Subventricular Zone of the Postnatal Mouse," *Environmental Pollution* 270 (2021): 116179, https://doi.org/10.1016/j.envpol.2020.116179.
29. Ya-Xing Gui et al., "Glyphosate Induced Cell Death through Apoptotic and Autophagic Mechanisms," *Neurotoxicology and Teratology* 34, no. 3 (2012): 344–49, https://doi.org/10.1016/j.ntt.2012.03.005.
30. Lucia O. Sampaio et al., "Heparins and Heparan Sulfates. Structure, Distribution and Protein Interactions," Chapter 1 in *Insights into Carbohydrate Structure and Biological Function*, ed. Hugo Verli (Kerala, India: Transworld Research Network, 2006), http://www.umc.br/_img/_noticias/755/artigo.pdf.
31. C. Pérez et al., "The Role of Heparan Sulfate Deficiency in Autistic Phenotype: Potential Involvement of Slit/Robo/srGAPs-Mediated Dendritic Spine Formation," *Neural Development* 11 (2016): 11, https://doi.org/10.1186/s13064-016-0066-x.
32. B. L. Pearson et al., "Heparan Sulfate Deficiency in Autistic Postmortem Brain Tissue from the Subventricular Zone of the Lateral Ventricles," *Behavioural Brain Research* 243 (2013): 138–45, https://doi.org/10.1016/j.bbr.2012.12.062.
33. F. Irie et al., "Autism-Like Socio-Communicative Deficits and Stereotypies in Mice Lacking Heparan Sulfate," *Proceedings of the National Academy of Sciences of the United States of America* 109, no. 13 (2012): 5052–56, https://doi.org/10.1073/pnas.1117881109.
34. P. J. Harrison et al., "Glutamate Receptors and Transporters in the Hippocampus in Schizophrenia," *Annals of the New York Academy of Sciences* 1003 (2003): 94–101, https://doi.org/10.1196/annals.1300.006.
35. Daiane Cattani et al., "Mechanisms Underlying the Neurotoxicity Induced by Glyphosate-Based Herbicide in Immature Rat Hippocampus: Involvement of Glutamate Excitotoxicity," *Toxicology* 320 (2014): 34–45, https://doi.org/10.1016/j.tox.2014.03.001.
36. L. A. Page et al., "*In Vivo* 1H-Magnetic Resonance Spectroscopy Study of Amygdala-Hippocampal and Parietal Regions in Autism," *American Journal of Psychiatry* 163 (2006): 2189–92, https://doi.org/10.1007/978-1-4939-7531-0_10.

37. C. Shimmura et al., "Alteration of Plasma Glutamate and Glutamine Levels in Children with High-Functioning Autism," *PLoS ONE* 6 (2011): e25340, https://doi.org/10.1371/journal.pone.0025340.
38. Daiane Cattani et al., "Mechanisms Underlying the Neurotoxicity Induced by Glyphosate-Based Herbicide in Immature Rat Hippocampus: Involvement of Glutamate Excitotoxicity," *Toxicology* 329 (2014): 34–45, https://doi.org/10.1016/j.tox.2014.03.001.
39. Daiane Cattani et al., "Developmental Exposure to Glyphosate-Based Herbicide and Depressive-Like Behavior in Adult Offspring: Implication of Glutamate Excitotoxicity and Oxidative Stress," *Toxicology* 387 (2017): 67–80, https://doi.org/10.1016/j.tox.2017.06.001.
40. Y. Bali et al., "Learning and Memory Impairments Associated to Acetylcholinesterase Inhibition and Oxidative Stress Following Glyphosate Based-Herbicide Exposure in Mice," *Toxicology* 415 (2019): 18–25, https://doi.org/10.1016/j.tox.2019.01.010.
41. L. Glusczak et al., "Effect of Glyphosate Herbicide on Acetylcholinesterase Activity and Metabolic and Hematological Parameters in Piava (*Leporinus obtusidens*)," *Environmental Toxicology and Pharmacology* 45 (2016): 41–44, https://doi.org/10.1016/j.ecoenv.2005.07.017.
42. A. Cieślińska et al., "Treating Autism Spectrum Disorder with Gluten-Free and Casein-Free Diet: The Underlying Microbiota-Gut-Brain Axis Mechanisms," *Journal of Clinical Immunology & Immunotherapy* 3 (2017): 009, https://doi.org/10.24966/CIIT-8844/100009.
43. Daniel C. Mathews et al., "Targeting the Glutamatergic System to Treat Major Depressive Disorder: Rationale and Progress to Date," Drugs 72, no. 10 (2012): 1313–33, https://doi.org/10.2165/11633130-000000000-00000.
44. Russell Blaylock, "Excitotoxins: The Taste That Kills," *Health Press* (1994).
45. Donald C. Rojas, "The Role of Glutamate and Its Receptors in Autism and the Use of Glutamate Receptor Antagonists in Treatment," *Journal of Neural Transmission* 121, no. 8 (2014): 891–905, https://doi.org/10.1007/s00702-014-1216-0.
46. Monika Krüger et al., "Field Investigations of Glyphosate in Urine of Danish Dairy Cows," *Journal of Environmental & Analytical Toxicology* 3 (2013): 5.
47. Lynn M. Kitchen et al., "Inhibition of δ-Aminolevulinic Acid Synthesis by Glyphosate," *Weed Science* 29, no 5 (1981): 571–77, https://doi.org/10.1017/S004317450006375X.
48. Yiting Zhang, "Decreased Brain Levels of Vitamin B_{12} in Aging, Autism and Schizophrenia," *PLoS One* 11, no. 1 (2016): e0146797, https://doi.org/10.1371/journal.pone.0146797.
49. Yiting Zhang et al., "Decreased Brain Levels of Vitamin B_{12} in Aging, Autism and Schizophrenia," *PLoS ONE* 11, no. 1 (2016): e0146797, https://doi.org/10.1371/journal.pone.0146797.

50. S. Jill James, "Metabolic Biomarkers of Increased Oxidative Stress and Impaired Methylation Capacity in Children with Autism," *American Journal of Clinical Nutrition* 80, no. 6 (2004): 1611–17, https://doi.org/10.1093/ajcn/80.6.1611.
51. K. L. Hung et al., "Cyanocobalamin, Vitamin B_{12}, Depresses Glutamate Release through Inhibition of Voltage-Dependent Ca2+ Influx in Rat Cerebrocortical Nerve Terminals (Synaptosomes)," *European Journal of Pharmacology* 602, nos. 2–3 (2009): 230–37, https://doi.org/10.1016/j.ejphar.2008.11.059.
52. Sally M. Pacholok and J. J. Stuart, *Could It Be B_{12}? An Epidemic of Misdiagnoses,* 2nd ed. (Fresno, CA: Quill Driver Books, 2011).
53. S. Seneff and G. Nigh, "Sulfate's Critical Role for Maintaining Exclusion Zone Water: Dietary Factors Leading to Deficiencies," *Water* 11 (2019): 22–42, https://doi.org/10.14294/WATER.2019.5.
54. S. Seneff and G. Nigh, "Sulfate's Critical Role," 22–42.
55. David A. Menassa et al., "Primary Olfactory Cortex in Autism and Epilepsy: Increased Glial Cells in Autism," *Brain Pathology* 27, no. 4 (2016): 437–48, https://doi.org/10.1111/bpa.12415.
56. Latifa S. Abdelli et al., "Propionic Acid Induces Gliosis and Neuro-inflammation through Modulation of PTEN/AKT Pathway in Autism Spectrum Disorder," *Scientific Reports* 9 (2019): 8824, https://doi.org/10.1038/s41598-019-45348-z.
57. S. R. Shultz et al., "Intracerebroventricular Injection of Propionic Acid, an Enteric Bacterial Metabolic End-Product, Impairs Social Behavior in the Rat: Implications for an Animal Model of Autism," *Neuropharmacology* 54 (2008): 901–11, https://doi.org/10.1016/j.neuropharm.2008.01.013.
58. Michael D. Innis, "Autism—An Immune Complex Disorder Following MMR?" *British Medical Journal* 2002, 325:419.
59. S. Toyoshima et al., "Methylmalonic Acid Inhibits Respiration in Rat Liver Mitochondria," *Journal of Nutrition* 125, no. 11 (1995): 2846–50, https://doi.org/10.1093/jn/125.11.2846.
60. Lamis Yehia et al., "Distinct Alterations in Tricarboxylic Acid Cycle Metabolites Associate with Cancer and Autism Phenotypes in Cowden Syndrome and Bannayan-Riley-Ruvalcaba Syndrome," *American Journal of Human Genetics* 105, no. 4 (2019): 813–21, https://doi.org/10.1016/j.ajhg.2019.09.004.
61. Nora Benachour and Gilles-Eric Séralini, "Glyphosate Formulations Induce Apoptosis and Necrosis in Human Umbilical, Embryonic, and Placental Cells," *Chemical Research in Toxicology* 22, no. 1 (2009): 97–105, https://doi.org/10.1021/tx800218n.
62. Christian W. Gruber, "Physiology of Invertebrate Oxytocin and Vasopressin Neuropeptides," *Experimental Physiology* 99 (2014): 55–56, https://doi.org/10.1113/expphysiol.2013.072561.
63. Karen J. Parker et al., "Intranasal Oxytocin Treatment for Social Deficits and Biomarkers of Response in Children with Autism," *Proceedings of the National*

Academy of Sciences U S A 114, no. 30 (2017): 8119–24, https://doi.org/10.1073/pnas.1705521114.

64. Srishti Shrestha et al., "Pesticide Use and Incident Hypothyroidism in Pesticide Applicators in the Agricultural Health Study," *Environmental Health Perspectives* 126, no. 9 (2018): 097008, https://doi.org/10.1289/EHP3194.
65. Janaina Sena de Souza et al., "Perinatal Exposure to Glyphosate-Based Herbicide Alters the Thyrotrophic Axis and Causes Thyroid Hormone Homeostasis Imbalance in Male Rats," *Toxicology* 377 (2017): 25–37, https://doi.org/10.1007/s00204-018-2236-6.
66. Fabiana Manservisi et al., "The Ramazzini Institute 13-week Pilot Study Glyphosate-Based Herbicides Administered at Human-equivalent Dose to Sprague Dawley Rats: Effects on Development and Endocrine System," *Environmental Health* 2019, 18, article 15, https://doi.org/10.1186/s12940-019-0453-y.
67. G. C. Román et al., "Association of Gestational Maternal Hypothyroxinemia and Increased Autism Risk," *Annals of Neurology* 74, no. 5 (2013): 733–42, https://doi.org/10.1002/ana.23976.

Chapter 10: Autoimmunity

1. Sean P. Keehan et al., "National Health Expenditure Projections, 2019–28: Expected Rebound in Prices Drives Rising Spending Growth," *Health Affairs* 39, no. 4 (2020): 704–14, https://www.doi.org/10.1377/hlthaff.2020.00094.
2. The Commonwealth Fund, "U.S. Health Care from a Global Perspective, 2019: Higher Spending, Worse Outcomes?" January 30, 2020, https://www.commonwealthfund.org/publications/issue-briefs/2020/jan/us-health-care-global-perspective-2019.
3. Centers for Disease Control and Prevention, "Health and Economic Costs of Chronic Diseases," November 17, 2020, https://www.cdc.gov/chronicdisease/about/costs/index.htm.
4. Gregg E. Dinse et al., "Increasing Prevalence of Antinuclear Antibodies in the United States," *Arthritis & Rheumatology* 72, no. 6 (2020): 1026–35, https://doi.org/10.1002/art.41214.
5. R. S. Gupta et al., "The Prevalence, Severity, and Distribution of Childhood Food Allergy in the United States," *Pediatrics* 128 (2011): e9–17, https://doi.org/10.1542/peds.2011-0204.
6. Kristen D. Jackson et al., "Trends in Allergic Conditions Among Children: United States, 1997–2011," Centers for Disease Control and Prevention, NCHS Data Brief No. 121, May 2013.
7. H. Okada et al., "The 'Hygiene Hypothesis' for Autoimmune and Allergic Diseases: an Update," *Clinical & Experimental Immunology* 160, no. 1 (2010): 1–9, https://doi.org/10.1111/j.1365-2249.2010.04139.x.

8. Michelle M. Stein et al., "Innate Immunity and Asthma Risk in Amish and Hutterite Farm Children," *New England Journal of Medicine* 375. no. 5 (2016) 411–21, https://doi.org/I: 10.1056/NEJMoa1508749.
9. Tamara Tuuminen and Kyösti Sakari Rinne, "Severe Sequelae to Mold-Related Illness as Demonstrated in Two Finnish Cohorts," *Frontiers in Immunology* 8 (2017 April): 382, https://doi.org/10.3389/fimmu.2017.00382.
10. Abi Berger, "Th1 and Th2 Responses: What Are They?" *British Medical Journal* 321, no. 7258 (2000): 424, https://doi.org/10.1136/bmj.321.7258.424.
11. Hongtao Liu et al., "TNF-alpha-Induced Apoptosis of Macrophages Following Inhibition of NF-kappaB: A Central Role for Disruption of Mitochondria," *Journal of Immunology* 172 (2004): 1907–15, https://doi.org/10.4049/jimmunol.172.3.1907.
12. Yanjing Xiao et al., "Structure-Activity Relationships of Fowlicidin-1, a Cathelicidin Antimicrobial Peptide in Chicken," *FEBS Journal* 273 (2006): 2581–93, https://doi.org/10.1111/j.1742-4658.2006.05261.x.
13. S. Lata et al., "AntiBP2: Improved Version of Antibacterial Peptide Prediction," *BMC Bioinformatics* 11, Suppl 1 (2010): S19, https://doi.org/10.1186/1471-2105-11-S1-S19.
14. B. Brodsky and A. V. Persikov, "Molecular Structure of the Collagen Triple Helix," *Advances in Protein Chemistry* 70 (2005): 301–39, https://doi.org/10.1016/S0065-3233(05)70009-7.
15. R. Sim et al., "C1q Binding and Complement Activation by Prions and Amyloids," *Immunobiology* 212 (2007): 355–62, https://doi.org/10.1016/j.imbio.2007.04.001.
16. Izma Abdul Zani et al., "Scavenger Receptor Structure and Function in Health and Disease," *Cells* 4 (2015): 178–201, https://doi.org/i: 10.3390/cells4020178.
17. S. Józefowski and L. Kobzik, "Scavenger Receptor A Mediates H2O2 Production and Suppression of IL-12 Release in Murine Macrophages," *Journal of Leukocyte Biology* 76, no. 5 (2004): 1066–74, https://doi.org/10.1189/jlb.0504270.
18. D. M. Steel et al., "The Major Acute Phase Reactants: C-Reactive Protein, Serum Amyloid P Component and Serum Amyloid A Protein," *Immunology Today* 15 (1994): 81–88, https://doi.org/10.1016/0167-5699(94)90138-4.
19. Aparamita Pandey et al., "Inflammatory Effects of Subacute Exposure of Roundup in Rat Liver and Adipose Tissue," *Dose Response* 17, no. 2 (2019): 1559325819843380, https://doi.org/10.1177/1559325819843380.
20. Asmita Pathak and Alok Agrawal, "Evolution of C-Reactive Protein," *Frontiers in Immunology* 10 (2019): 943, https://doi.org/10.3389/fimmu.2019.00943.
21. Alok Agrawal et al., "Pattern Recognition by Pentraxins," *Advances in Experimental Medicine and Biology* 653 (2009): 98–116, https://doi.org/10.1007/978-1-4419-0901-5_7.

22. Sarah E. Clark and Jeffrey N. Weiser, "Microbial Modulation of Host Immunity with the Small Molecule Phosphorylcholine," *Infection and Immunity* 81, no. 2 (2013): 392–401, https://doi.org/10.1128/IAI.01168-12.
23. Debra Gershov et al., "C-Reactive Protein Binds to Apoptotic Cells, Protects the Cells from Assembly of the Terminal Complement Components, and Sustains an Anti-inflammatory Innate Immune Response: Implications for Systemic Autoimmunity," *Journal of Experimental Medicine* 192, no. 9 (2000): 1353–63, https://doi.org/10.1084/jem.192.9.1353.
24. Debra Gershov et al., "C-Reactive Protein Binds to Apoptotic Cells, Protects the Cells from Assembly of the Terminal Complement Components, and Sustains an Anti-inflammatory Innate Immune Response: Implications for Systemic Autoimmunity," *Journal of Experimental Medicine* 192, no. 9 (2000): 1353–63, https://doi.org/10.1084/jem.192.9.1353.
25. T. W. Faust et al., "Neurotoxic Lupus Autoantibodies Alter Brain Function through Two Distinct Mechanisms," *Proceedings of the National Academy of Sciences of the United States of America* 107, no. 43 (2010): 18569–74, https://doi.org/10.1073/pnas.1006980107.
26. Y. Li et al., "Monocyte and Macrophage Abnormalities in Systemic Lupus Erythematosus," *Archivum Immunologiae et Therapiae Experimentalis (Warsz)* 58, no. 5 (2010): 355–64, https://doi.org/q0.1007/s00005-010-0093-y.
27. Shaye Kivity et al., "Neuropsychiatric Lupus: A Mosaic of Clinical Presentations," *BMC Medicine* 13 (2015): 43.
28. Hui Gao et al., "Activation of the *N*-methyl-D-aspartate Receptor Is Involved in Glyphosate-Induced Renal Proximal Tubule Cell Apoptosis," *Journal of Applied Toxicology* (2019): 1–12, https://doi.org/10.1002/jat.3795; Daiane Cattani et al., "Mechanisms Underlying the Neurotoxicity Induced by Glyphosate-Based Herbicide in Immature Rat Hippocampus: Involvement of Glutamate Excitotoxicity," *Toxicology* 320 (2014): 34–45, https://doi.org/10.1016/j.tox.2014.03.001; Daiane Cattani et al., "Developmental Exposure to Glyphosate-Based Herbicide and Depressive-Like Behavior in Adult Offspring: Implication of Glutamate Excitotoxicity and Oxidative Stress," *Toxicology* 387 (2017): 67–80, https://doi.org/10.1016/j.tox.2017.06.001.
29. D .A. Fraser and A. J. Tenner, "Innate Immune Proteins C1q and Mannan-Binding Lectin Enhance Clearance of Atherogenic Lipoproteins by Human Monocytes and Macrophages," *Journal of Immunology* 185 (2010): 3932–39, https://doi.org/10.4049/jimmunol.1002080.
30. T. Kawashima et al., "Impact of Ultraviolet Irradiation on Expression of SSA/Ro Autoantigenic Polypeptides in Transformed Human Epidermal Keratinocytes," *Lupus* 3, no. 6 (1994): 493–500, https://doi.org/10.1177/096120339400300612.

31. Rosanne A. van Schaarenburg et al., "C1q Deficiency and Neuropsychiatric Systemic Lupus Erythematosus," *Frontiers in Immunology* 7 (2016): 647, https://doi.org/10.3389/fimmu.2016.00647.
32. C. Esposito et al., "New Therapeutic Strategies for Coeliac Disease: Tissue Transglutaminase as a Target," *Current Medicinal Chemistry* 14, no. 24 (2007): 2572–80, https://doi.org/10.2174/092986707782023343; N. A. Molodecky et al., "Increasing Incidence and Prevalence of the Inflammatory Bowel Diseases with Time, Based on Systematic Review," *Gastroenterology* 142, no. 1 (2012): 46–54, https://doi.org/10.1053/j.gastro.2011.10.001.
33. A. Rubio-Tapia et al., "Increased Prevalence and Mortality in Undiagnosed Celiac Disease," *Gastroenterology* 137, no. 1 (2009): 88–93, https://doi.org/10.1053/j.gastro.2009.03.059.
34. George Janssen et al., "Ineffective Degradation of Immunogenic Gluten Epitopes by Currently Available Digestive Enzyme Supplements," *PLoS ONE* 10, no. 6 (2015): e0128065, https://doi.org/10.1371/journal.pone.0128065.
35. Kate Beaudoin and Darryn S. Willoughby, "The Role of the Gluten-Derived Peptide Gliadin in Celiac Disease," *Journal of Nutritional Health & Food Engineering* 1, no. 6 (2014): 229–32, https://doi.org/10.15406/jnhfe.2014.01.00036.
36. T. Byun et al., "Synergistic Action of an X-Prolyl Dipeptidyl Aminopeptidase and a Non-Specific Aminopeptidase in Protein Hydrolysis," *Journal of Agricultural and Food Chemistry* 49, no. 4 (2001): 2061–63, https://doi.org/10.1021/jf001091m.
37. Fabienne Morel et al., "The Prolyl Aminopeptidase from *Lactobacillus delbrueckii* subsp. *bulgaricus* Belongs to the α/β Hydrolase Fold Family," *Biochimica et Biophysica Acta* 1429 (1999): 501–5, https://doi.org/10.1016/s0167-4838(98)00264-7.
38. Alberto Caminero et al., "Duodenal Bacterial Proteolytic Activity Determines Sensitivity to Dietary Antigen through Protease-Activated Receptor-2," *Nature Communications* 10 (2019): 1198, https://doi.org/10.1038/s41467-019-09037-9.
39. K. G. Kerr and A. M. Snelling, "*Pseudomonas aeruginosa*: A Formidable and Ever-Present Adversary," *Journal of Hospital Infection* 73, no. 4 (2009): P338–344, https://doi.org/10.1016/j.jhin.2009.04.020.
40. Maria Laura Cupi et al., "Defective Expression of Scavenger Receptors in Celiac Disease Mucosa," *PLoS One* 9, no. 6 (2014): e100980, https://doi.org/10.1007/s00109-004-0623-3; Michele Boniotto et al., "Evidence of a Correlation between Mannose Binding Lectin and Celiac Disease: A Model for Other Autoimmune Diseases," *Journal of Molecular Medicine* 83, no. 4 (2005): 308–15, https://doi.org/10.1007/s00109-004-0623-3.
41. Jorge Escobedo-de la Peña et al., "Hypertension, Diabetes and Obesity, Major Risk Factors for Death in Patients With COVID-19 in Mexico," *Archives of*

Medical Research, December 16, 2020 [Epub ahead of print], https://doi.org/10.1016/j.arcmed.2020.12.002.

42. Nancy L. Swanson et al., "Genetically Engineered Crops, Glyphosate and the Deterioration of Health in the United States of America," *Journal of Organic Systems* 9 (2014): 6–37.
43. Katarina Zimmer, "The Immune Hallmarks of Severe COVID-19," September 16, 2020, https://www.the-scientist.com/news-opinion/the-immune-hallmarks-of-severe-covid-19-67937; Shintaro Hojyo et al., "How COVID-19 Induces Cytokine Storm with High Mortality," *Inflammation and Regeneration* 40 (2020): 37, https://doi.org/10.1186/s41232-020-00146-3.
44. Mariam Ahmed Saad et al., "Covid-19 and Autoimmune Diseases: A Systematic Review of Reported Cases," *Current Rheumatology Reviews* 2020 Oct 29 [Epub ahead of print], https://doi.org/10.2174/1573397116666201029155856.
45. Jonas Blomberg et al., "Infection Elicited Autoimmunity and Myalgic Encephalomyelitis/Chronic Fatigue Syndrome: An Explanatory Model," *Frontiers in Immunology* 9 (2018): 22, https://doi.org/10.3389/fimmu.2018.0022.
46. H. D. Bremer, N. Landegren, R. Sjöberg et al., "ILF2 and ILF3 Are Autoantigens in Canine Systemic Autoimmune Disease," *Scientific Reports* 8, (2018): 4852, https://doi.org/10.1038/s41598-018-23034-w.
47. Heng L. Tham et al., "Autoimmune Diseases Affecting Skin Melanocytes in Dogs, Cats and Horses: Vitiligo and the Uveodermatological Syndrome: A Comprehensive Review," *BMC Veterinary Research* 15 (2019): 251, https:/doi.org/10.1186/s12917-019-2003-9.

Chapter 11: Reboot Today for a Healthy Tomorrow

1. Lisa Zimmermann et al., "Benchmarking the In Vitro Toxicity and Chemical Composition of Plastic Consumer Products," *Environmental Science & Technology* 53, no. 19 (2019) 11467–77, https://doi.org/10.1021/acs.est.9b02293.
2. Matjec Mikulic, "Global Pharmaceutical Industry—Statistics & Facts," Statista (November 5, 2008): https://www.statista.com/topics/1764/global-pharmaceutical-industry.
3. "Autoimmune Disease Treatment Market a $108 Billion Industry Much Bigger Than Cancer and Heart Disease—Forecast 2018–2023," Medgadget, updated October 5, 2018, https://www.medgadget.com/2018/10/autoimmune-disease-treatment-market-a-108-billion-industry-much-bigger-than-cancer-and-heart-disease-forecast-2018-2023.html.
4. Daniel McDonald et al., "American Gut: An Open Platform for Citizen Science Microbiome Research," *mSystems* 3, no. 3 (2018): e00031–18, https://doi.org/OI: 10.1128/mSystems.00031-18.
5. A. Tarozzi et al., "Sulforaphane as an Inducer of Glutathione Prevents Oxidative Stress-Induced Cell Death in a Dopaminergic-Like Neuroblastoma Cell Line,"

Journal of Neurochemistry 11, no. 5 (2009): 1161–71, https://doi.org/10.1111/j.1471-4159.2009.06394.x.

6. T. W. Sedlak et al., "Sulforaphane Augments Glutathione and Influences Brain Metabolites in Human Subjects: A Clinical Pilot Study," *Molecular Neuropsychiatry* 3 (2017): 214–22, https://doi.org/10.1159/000487639.
7. Ruhi Turkmen et al., "Antioxidant and Cytoprotective Effects of *N*-acetylcysteine against Subchronic Oral Glyphosate-Based Herbicide-Induced Oxidative Stress in Rats," *Environmental Science and Pollution Research International* 26, no. 11 (2019): 11427–37, https://doi.org/10.1007/s11356-019-04585-5.
8. Anke Bongers et al., "Prebiotics and the Bioavailability of Minerals and Trace Elements," *Reviews International* 19, no. 4 (2003): 397–422, https://doi.org/10.1081/FRI-120025482.
9. M. Ferrer et al., "Influence of Prebiotics, Probiotics and Protein Ingredients on Mycotoxin Bioaccessibility," *Food & Function* 6 (2015): 987–94, https://doi.org/10.1039/c4fo01140f.
10. Justin L. Carlson et al., "Health Effects and Sources of Prebiotic Dietary Fiber," *Current Developments in Nutrition* 2, no. 3 (2018): nzy005, https://doi.org/10.1093/cdn/nzy005.
11. Y. A. Dound et al., "The Effect of Probiotic Bacillus subtilis HU58 on Immune Function in Healthy Human," *The Indian Practitioner* 70, no. 9 (2017): 15–20.
12. Cindy Duysburgh et al., "A Synbiotic Concept Containing Spore-Forming Bacillus Strains and a Prebiotic T Fiber Blend Consistently Enhanced Metabolic Activity by Modulation of the Gut Microbiome *in Vitro*," *International Journal of Pharmaceutics* X 1 (2019): 100021, https://doi.org/10.1016/j.ijpx.2019.100021.
13. Khalid AlFaleh and Jasim Anabrees, "Probiotics for Prevention of Necrotizing Enterocolitis in Preterm Infants," *Cochrane Database Systematic Reviews* 10, no. 4 (2014): CD005496, https://doi.org/10.1002/14651858.CD005496.pub4; Guillermo Bernaola Aponte et al., "Probiotics for Treating Persistent Diarrhoea in Children," *Cochrane Database Systematic Reviews* 10, no. 11 (2010): CD007401, https://doi.org/10.1002/14651858.CD007401.pub2; Joshua Z. Goldenberg et al., "Probiotics for the Prevention of Clostridium difficile—Associated Diarrhea in Adults and Children," *Cochrane Database Systematic Reviews* 12, no. 12 (2017): CD006095, https://doi.org/10.1002/14651858.CD006095.pub4.
14. James A. McCubrey et al., "Effects of Resveratrol, Curcumin, Berberine and Other Nutraceuticals on Aging, Cancer Development, Cancer Stem Cells and MicroRNAs," *Aging (Albany NY)* 9, no. 6 (2017): 1477–536, https://doi.org/10.18632/aging.101250.
15. Haneen Amawi et al., "Polyphenolic Nutrients in Cancer Chemoprevention and Metastasis: Role of the Epithelial-to-Mesenchymal (EMT) Pathway," *Nutrients* 9 (2017): 911, https://doi.org/10.3390/nu9080911.

16. A. Kawasaki et al., "The Taurine Content of Japanese Seaweed," *Advances in Experimental Medicine and Biology* 975, Pt 2 (2017): 1105–12.
17. Sally Fallon Morell, "Dissecting Those New Fake Burgers," *Wise Traditions in Food, Farming, and the Healing Arts* 20, no. 3 (2019): 38–45, https://doi.org/10.1007/978-94-024-1079-2_88.
18. Eat Fit Go, "Bunless Beyond Burger—Vacuum Packed," last accessed December 3, 2020, https://www.eatfitgo.com/products/bunless-beyond-burger-vacuum-packed.
19. Zen Honeycutt, "GMO Impossible Burger Positive for Carcinogenic Glyphosate," Moms Across America, May 16, 2019, last accessed Nov. 6, 2019, https://www.momsacrossamerica.com/gmo impossible burger positive for carcinogenic glyphosate.
20. David G. Hoel and Frank R. de Gruijl, "Sun Exposure Public Health Directives," *International Journal of Environmental Research and Public Health* 15 (2018): 2794, https://doi.org/10.3390/ijerph15122794.
21. P. G. Lindqvist et al., "Avoidance of Sun Exposure is a Risk Factor for All-Cause Mortality: Results from the Melanoma in Southern Sweden Cohort," *Journal of Internal Medicine* 276, no. 1 (2014): 77–86, https://doi.org/10.1111/joim.12251.
22. Dr. Joseph Mercola, "The Fourth Phase of Water—What You Don't Know About Water, and Really Should," August 18, 2013, https://articles.mercola.com/sites/articles/archive/2013/08/18/exclusion-zone-water.aspx; Xian He et al., "Effect of Spin Polarization on the Exclusion Zone of Water," *Journal of Physical Chemistry* B 122, no. 36 (2018): 8493–502, https://doi.org/10.1021/acs.jpcb.8b04118.
23. Stephanie Seneff et al., "Is Endothelial Nitric Oxide Synthase a Moonlighting Protein Whose Day Job Is Cholesterol Sulfate Synthesis? Implications for Cholesterol Transport, Diabetes and Cardiovascular Disease," *Entropy* 14 (2012): 2492–530.
24. Lauren A. Burt et al., "Effect of High-Dose Vitamin D Supplementation on Volumetric Bone Density and Bone Strength: A Randomized Clinical Trial," *Journal of the American Medical Association* 322, no. 8 (2019): 736–45, https://doi.org/10.1001/jama.2019.11889.
25. Vivek G. Patwardhan et al., "Randomized Control Trial Assessing Impact of Increased Sunlight Exposure versus Vitamin D Supplementation on Lipid Profile in Indian Vitamin D Deficient Men," *Indian Journal of Endocrinology and Metabolism* 21, no. 3 (2017): 393, https://pubmed.ncbi.nlm.nih.gov/28553593/.
26. Natalie H. Matthews et al., "Chapter 1: Epidemiology of Melanoma," in Cutaneous Melanoma: Etiology and Therapy, W. H. Ward and J. M. Farma, eds. (Brisbane, Australia: Codon Publications, 2017).
27. Weisheng Lin et al., "Toxicity of Nano- and Micro-Sized ZnO Particles in Human Lung Epithelial Cells," *Journal of Nanoparticle Research* 11 (2009): 25–39, https://doi.org/10.1007/s11051-008-9419-7.

28. K. M. Hanson et al., "Sunscreen Enhancement of UV-Induced Reactive Oxygen Species in the Skin," *Free Radical Biology & Medicine* 41 (2006): 1205–12, https://doi.org/10.1016/j.freeradbiomed.2006.06.011.
29. S. Nicholson and C. Exley, "Aluminum: A Potential Pro-oxidant in Sunscreens/Sunblocks?" *Free Radical Biology & Medicine* 43, no. 8 (2007): 1216–17, https://doi.org/10.1016/j.freeradbiomed.2007.07.010.
30. E. H. Winfried et al., "Does Soil Contribute to the Human Gut Microbiome?" *Microorganisms* 7 (2019): 287, https://doi.org/10.3390/microorganisms7090287.
31. Gaétan Chevalier et al, "Earthing: Health Implications of Reconnecting the Human Body to the Earth's Surface Electrons," *Journal of Environmental and Public Health* (2012): Article ID 291541, https://doi.org/10.1155/2012/291541.
32. Martin L. Pall, "Electromagnetic Fields Act via Activation of Voltage-Gated Calcium Channels to Produce Beneficial or Adverse Effects," *Journal of Cellular and Molecular Medicine* 17, no. 8 (2013): 958–65, https://doi.org/10.1111/jcmm.12088.
33. P. L. Eberbach and L. A. Douglas, "Persistence of Glyphosate in a Sandy Loam," *Soil Biology and Biochemistry* 15, no. 4 (1983): 485–87, https://doi.org/10.1016/0038-0717(83)90016-0.
34. Jingwen Xu et al., "Glyphosate Contamination in Grains and Foods: An Overview," *Food Control* 106 (2019): 106710, https://doi.org/10.1016/j.foodcont.2019.106710; Paul Sprankle et al., "Adsorption, Mobility, and Microbial Degradation of Glyphosate in the Soil," Weed Science 23, no. 3 (1975): 229–34.
35. Hui Zhan et al., "Recent Advances in Glyphosate Biodegradation," *Applied Microbiology and Biotechnology* 102, no. 12 (2018): 5033–43, https://doi.org/10.1007/s00253-018-9035-0.
36. X. M. Yu et al., "Glyphosate Biodegradation and Potential Soil Bioremediation by *Bacillus subtilis* Strain Bs-15," *Genetics and Molecular Research* 14, no. 4 (2015): 14717–30, https://doi.org/10.4238/2015.November.18.37.
37. A. N. Moneke et al., "Biodegradation of Glyphosate Herbicide *in Vitro* Using Bacterial Isolates from Four Rice Fields," *African Journal of Biotehcnology* 9, no. 26 (2010): 4067–74.
38. Geoffrey Davies et al., "Humic Acids: Marvelous Products of Soil Chemistry," *Journal of Chemical Education* 78, no. 12 (2001): 1609, https://doi.org/10.1021/ed078p1609.
39. BioAg Europe, "Glyphosate and the Neutralizing Effect of Humic Acid," https://www.bioag.eu/en/glyphosate-and-the-neutralizing-effect-of-humic-acid/.
40. Monika Krüger et al., "Glyphosate Suppresses the Antagonistic Effect of *Enterococcus* spp. on *Clostridium botulinum*, *Anaerobe* 20 (2013): 74–78, https://doi.org/10.1016/j.anaerobe.2013.01.005; Henning Gerlach et al., "Oral Application of Charcoal and Humic Acids to Dairy Cows Influences Clostridium botulinum

Blood Serum Antibody Level and Glyphosate Excretion in Urine," *Journal of Clinical Toxicology* 4 (2014): 2; Awad A. Shehata et al., "Neutralization of the Antimicrobial Effect of Glyphosate by Humic Acid *in Vitro*," *Chemosphere* 104 (2014): 258–61, https://doi.org/10.1016/j.chemosphere.2013.10.064; Monika Krüger et al., "Field Investigations of Glyphosate in Urine of Danish Dairy Cows," *Journal of Environmental and Analytical Toxicology* 3 (2013): 5, http://dx.doi.org/10.4172/2161-0525.1000186.

41. Henning Gerlach et al., "Oral Application of Charcoal and Humic acids to Dairy Cows Influences *Clostridium botulinum* Blood Serum Antibody Level and Glyphosate Excretion in Urine," *Journal of Clinical Toxicology* 4 (2014): 2.
42. Awad A. Shehata et al., "Neutralization of the Antimicrobial Effect of Glyphosate by Humic Acid *in Vitro*," *Chemosphere* 104 (2014): 258–61, https://doi.org/10.1016/j.chemosphere.2013.10.064.
43. Olga V. Koroleva et al., "The Role of White-rot Fungi in Herbicide Transformation," Chapter 9 in Herbicides, Physiology of Action and Safety (London: Intech, 2015), http://dx.doi.org/10.5772/61623.
44. Naomi Farragher, "Degradation of Pesticides by the Ligninolytic Enzyme Laccase," master's thesis, Swedish University of Agricultural Sciences, 2013.
45. Leticia Pizzul et al., "Degradation of Glyphosate and Other Pesticides by Ligninolytic Enzymes," *Biodegradation* 20 (2009): 751, https://doi.org/10.1007/s10532-009-9263-1.
46. Naomi Farragher, "Degradation of Pesticides by the Ligninolytic Enzyme Laccase," master's thesis, Swedish University of Agricultural Sciences, 2013.
47. United States Environmental Protection Agency, "National Primary Drinking Water Regulations," https://www.epa.gov/ground-water-and-drinking-water/national-primary-drinking-water-regulations#Organic.
48. Thomas F. Speth, "Glyphosate Removal from Drinking Water," *Journal of Environmental Engineering* 119, no. 6 (1993): 1139–57.
49. Jörgen Jönsson et al., "Removal and Degradation of Glyphosate in Water Treatment: A Review," *AQUA* 62, no. 7 (2013): 395–408, https://doi.org/10.2166/aqua.2013.080.

Index

Note: Page numbers followed by *t* refer to tables.